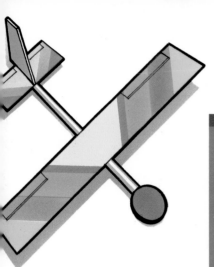

101 PHYSICS TRICKS

TERRY CASH

Fun Experiments with Everyday Materials

Sterling Publishing Co., Inc.   New York

CONTENTS

Library of Congress Cataloging-in-Publication Data

Cash, Terry.
    [Fun with physics]
    101 Physics tricks : fun experiments with everyday materials /
Terry Cash.
        p.   cm.
    "Originally published in Great Britain . . . under the title Fun
with physics, © 1991"—T.p. verso.
    Includes index.
    Summary: Provides information on such topics as gravity, friction,
sound, magnetism, light, heat, and energy and includes activities
and experiments illustrating various principles.
    ISBN 0-8069-8786-3
    1. Physics—Juvenile literature. 2. Physics—Experiments—
Juvenile literature. 3. Scientific recreations—Juvenile
literature. [1. Physics. 2. Physics—Experiments.
3. Experiments. 4. Scientific recreations.] I. Title. II. Title:
One hundred one physics tricks. III. Title: One hundred and one
physics tricks.
QC25.C37   1992
530'.078—dc20   92-21859
                                                            CIP
                                                            AC

10 9 8 7 6 5 4 3 2 1

First paperback edition published in 1993 by
Sterling Publishing Company, Inc.
387 Park Avenue South, New York, N.Y. 10016
Originally published in Great Britain by
Simon & Schuster Young Books under the title
Fun with Physics © 1991 by Terry Cash
Illustrations © 1991 by Simon & Schuster Young Books
Distributed in Canada by Sterling Publishing
c/o Canadian Manda Group. P.O. Box 920, Station U
Toronto, Ontario, Canada M8Z 5P9
Printed and bound in Hong Kong
All rights reserved
Sterling ISBN 0-8069 8786-3 Trade
            ISBN 0-8069-8785-5 Paper

1
PUFF
SQUEEZE
BANG
BLOW

THE SCIENCE AND TECHNOLOGY OF FORCES, AIR AND SOUND

GRAVITY

It is a fact of life, drop your favorite toy and it will crash to the ground, knock over a milk carton and the milk will pour all over the floor. Even if you throw a ball as hard as you can into the air, it will always fall back to the ground again. But why? In 1684 an English scientist, Sir Isaac Newton, explained his theory of gravity. We believe that as he was sitting in his father's orchard, he watched an apple fall. This set him thinking, why does everything fall to the ground?

Newton's theory
Newton showed that all things attract one another, but very large objects like planets and stars have enormous gravitational force. Everything on Earth is pulled by gravity towards its center. This explains why rivers flow downhill to reach their lowest level. Even light objects like hot-air balloons eventually float back to Earth again.

Getting away from it all

If a rocket is fast enough, it can break free of Earth's gravitational grip and fly into space. The very high speed needed is called the escape velocity (about 25,000 miles per hour).

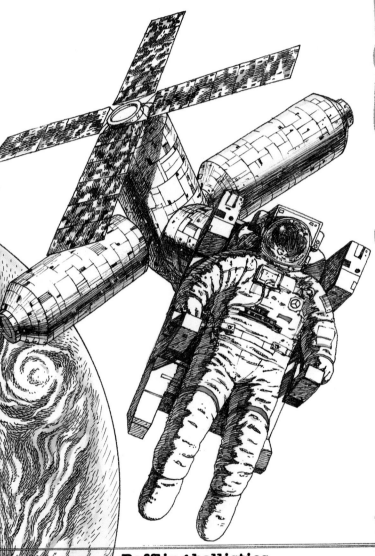

Baffling ballistics

When a gun is fired, the bullet is pushed at high speed. As it flies through the air it is pulled downwards by gravity until it hits the ground. If a second bullet is dropped at the same instant as the gun is fired, and the gun is held level with the ground, both bullets hit the ground together. The speed of the fired bullet makes no difference to the pull of gravity.

PRINCIPLES OF GRAVITY

The clink-clunk test

Choose several objects, like a marble and tennis ball, and guess which one will fall the fastest. Place a metal tray on the ground and hold two objects high above it. Let them go at the same time and see which one hits the tray first. Always repeat your experiments. What do you discover? (See page 8.)

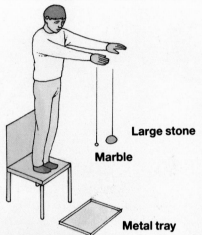

Large stone

Marble

Metal tray

Man on the moon

An astronaut loses weight as he travels away from Earth because there is less pull of gravity. Scientists use two methods of weighing. Spring balances measure the pull of gravity on an object. This is called weight and is measured in pounds. A beam balance measures the mass of an object by comparing it with standard masses called Newtons. The Moon is smaller and has less gravity, so people weigh about one sixth of their weight on Earth, but their mass stays the same.

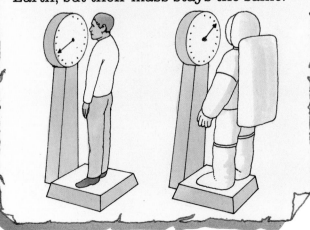

PUTTING THE BRAKES ON

If you tried the clink-clunk test, you found that all objects fall to the ground at the same speed and hit at the same time. The only difference is if the object is so light the air is able to hold it up. Hold a sheet of paper above your head and watch it float to the floor. The same sheet of paper rolled into a tight ball falls rapidly to the ground. The reason that the flat sheet floats down is because of its high wind resistance. Wind resistance depends on the surface area and mass of an object. The flat sheet has a large area that can be supported by the air to cushion its fall. When the sheet is screwed into a tight ball, its mass stays the same but there is much less area for the air to support — so it drops like a stone.

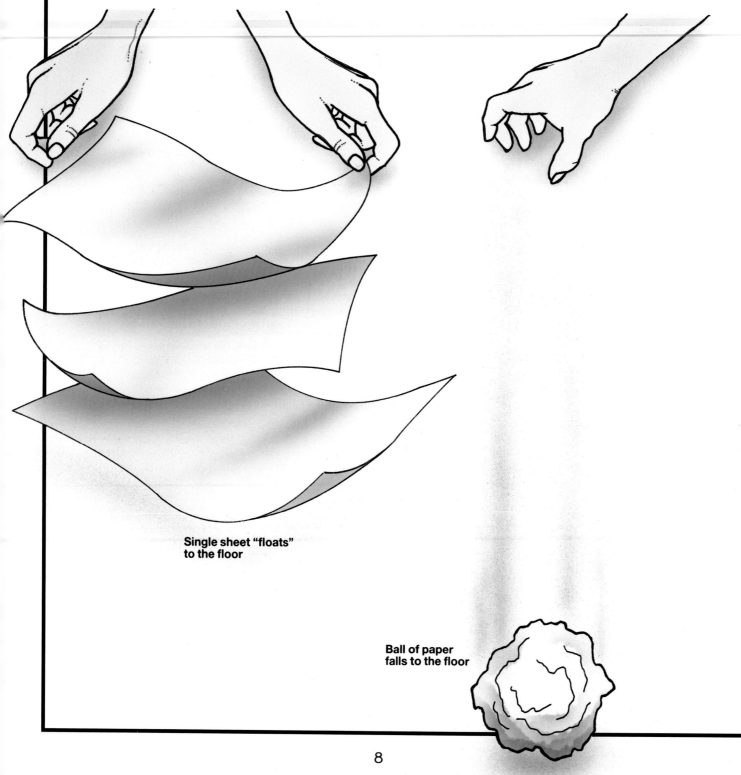

Single sheet "floats" to the floor

Ball of paper falls to the floor

MAKE A PARACHUTE

Cut a large circle from a plastic bag. Cut six pieces of thread or thin string. Make them all the same length (about one and a half times the diameter of the circle). Tape each string to the edge of the circle and tie the other ends together around a small toy. Throw your parachute high into the air and watch it float to the ground.

Experiment with different sizes of parachute and smaller or larger toys to find which parachute gives the longest flight and the softest landing.

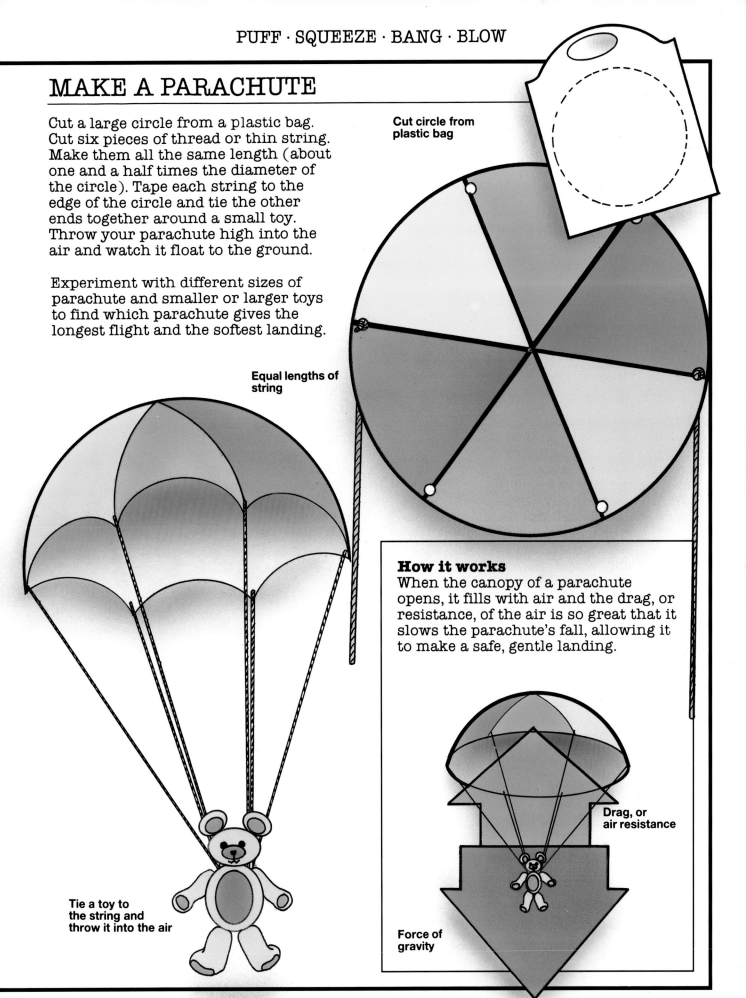

Cut circle from plastic bag

Equal lengths of string

How it works
When the canopy of a parachute opens, it fills with air and the drag, or resistance, of the air is so great that it slows the parachute's fall, allowing it to make a safe, gentle landing.

Drag, or air resistance

Tie a toy to the string and throw it into the air

Force of gravity

FRICTION

When two surfaces rub against each other the force of friction causes them to "stick" together. Friction stops our shoes from slipping on the pavement. Friction between brake blocks and wheel rims slows down bicycles. Friction lets us unscrew bottle tops. But friction is also destructive. The moving parts of engines rub together and eventually wear out. Clothes and shoes wear thin when they rub against the floor and furniture. Even paintwork on aircraft is scrubbed off by the rush of air.

Fire from friction

Friction makes heat. If you rub your hands together they soon become warm. Prehistoric people found that they could spin a dry twig on a wooden block and produce fire. As the point of the twig rubbed against the block, it got hotter until it started to burn.

Skating along

Ice is very slippery — it has very low friction. Expert skaters can move very fast across the ice because friction is reduced and does not hold them back.

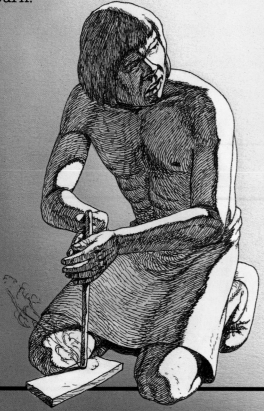

A safer Earth

The Moon and many of the planets are deeply pitted and cratered where they have been struck by meteorites. Only a few have ever hit Earth because the layer of air around our planet, the atmosphere, acts like a protective blanket. Small meteorites hit the atmosphere at high speed and the friction caused by the air rubbing against them is so great that they burn up in a flash of light.

PRINCIPLES OF FRICTION

Lubrication

Unscrewing a bottle top with clean, dry hands is no problem, but try again with soapy hands. The soapy layer is slippery and it is impossible to get a good grip. Friction is reduced. This is not always helpful, but making things more slippery can be very useful and is called lubrication.

No soap   **With soap**

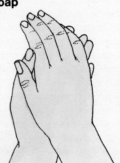 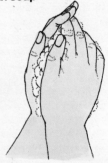

Running smooth

If engine parts are allowed to rub against one another, the metal surfaces get very hot and soon wear out. But if oil is poured over the moving parts of an engine, it forms a thin layer between the metal surfaces, so that they don't actually touch. This lets them move smoothly and reduces wear.

NO OIL
high friction great heat

WITH OIL
little friction less heat

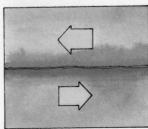 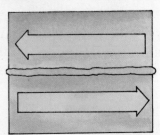

The rough edges of metal engine parts greatly magnified

TESTING "GRIPABILITY"

Find a number of different objects such as an eraser, a coin, a bottle cap and even an ice cube. Place them at one end of a large, smooth board. Gently and slowly lift the end of the board, watching closely to see which object begins to slide first. Observe the angle at which each moves. Which object has a low friction and which has the greatest grip? Does it alter the results if you first sprinkle the board with water or put polish onto the surface?

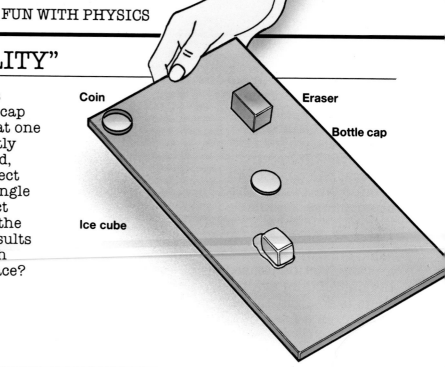

Coin

Eraser

Bottle cap

Ice cube

Egyptian engineering

Tie a piece of string to a shoebox filled with toys and put it on a table. Try pulling it across the table. Is it hard to move? Put a lump of clay on the end of the string. How much do you need to make the box move? The Egyptians used wooden rollers to move the stones needed for their pyramids, but do they make it easier? Put straws or pencils under the box. How much clay do you need to move the box? The rollers help to reduce friction, but does the size of roller matter? Try different-sized rollers.

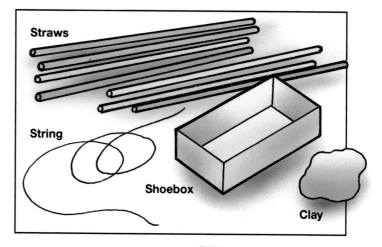

Straws

String

Shoebox

Clay

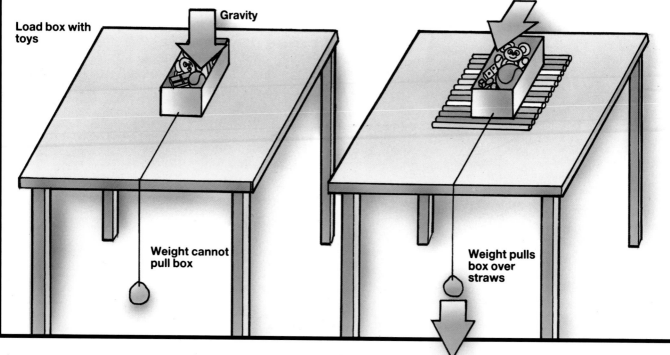

Load box with toys

Gravity

Weight cannot pull box

Weight pulls box over straws

Road safety

Put a motorized toy car at the bottom of a large shiny board. Set the board at the steepest angle that your car can just manage to go up. Now wrap cellophane tape around the tires and try again. Can your car still climb the hill? The smooth, slippery surface of the tape makes the wheels spin without getting a good grip. This is why it is so dangerous for cars and trucks to be driven with tires that have worn smooth. Friction is reduced.

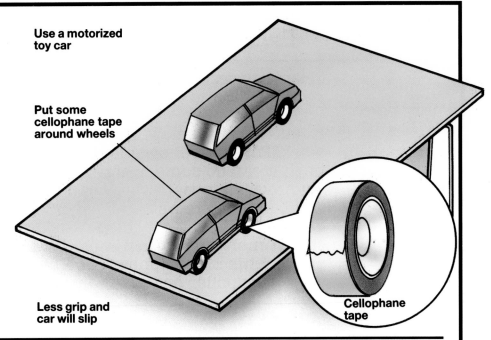

Use a motorized toy car

Put some cellophane tape around wheels

Less grip and car will slip

Cellophane tape

Aerodynamics

Which car will get to the bottom of the slope first? The faster a car goes, the harder it is hit by the air, which holds it back. If the car is badly designed, with large square panels, there is great air resistance. But if the car has a smooth, streamlined shape, it can cut easily through the air, travelling faster and more economically.

Tape

Toy cars

Scissors

Hairdrier

Streamlined

Tape below

Tape

Great air resistance

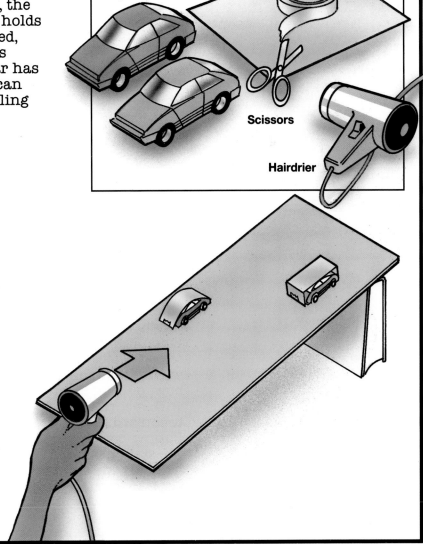

ELASTIC POWER

Scientists describe two forms of energy. One is called kinetic energy, the other is called potential energy. When something is moving it has kinetic energy. It is doing some form of work. Potential energy is stored in an object until it can do some useful work. For example, potential energy can be stored in an elastic band by stretching or twisting it. When the band is released it gives out its stored energy by returning to its original shape. This kinetic energy can be used to turn a propeller or to launch a missile.

Twisted strings
The Romans used a mangonel to hurl rocks. As its arm was pulled down, the arm twisted ropes at its base. When the arm was released, the energy stored in the twisted ropes flung the rocks a long way.

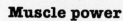

Bending wood
Energy is stored in wood when it is bent. A bowman would select a strong, springy branch for his bow, pulling it into shape with a cord made from tough animal sinew. Pulling back an arrow on the string, meant both the bow and the string were bent, ready to spring back, so the arrow was fired at great speed.

Muscle power
Muscles can only work by pulling. They need to be stretched before they can pull again. Many of the muscles in our bodies work in pairs. While one is pulling, the other is being stretched ready to work.

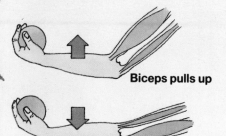

Biceps pulls up

Triceps pulls down

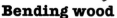

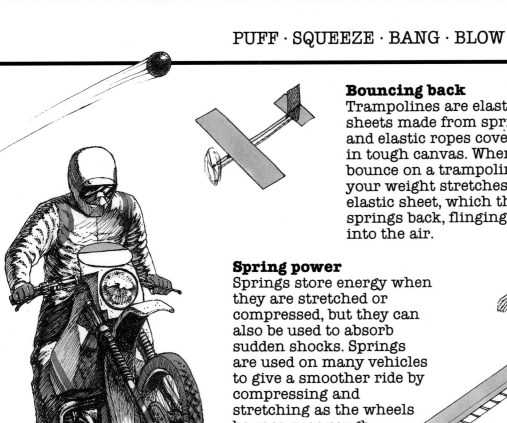

Bouncing back

Trampolines are elastic sheets made from springs and elastic ropes covered in tough canvas. When you bounce on a trampoline, your weight stretches the elastic sheet, which then springs back, flinging you into the air.

Spring power

Springs store energy when they are stretched or compressed, but they can also be used to absorb sudden shocks. Springs are used on many vehicles to give a smoother ride by compressing and stretching as the wheels bounce over rough ground.

PRINCIPLES OF ELASTICITY

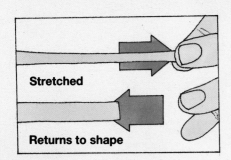

Stretched

Returns to shape

Rubber

Rubber is a natural elastic material. The atoms that make up rubber form long chains. You can stretch rubber out of shape but when you let go of it, the long chains pull the rubber back into its original shape.

Make a force meter

The force of gravity pulling on an object is measured with a force meter. Ask an adult to knock a nail into a board and hang a strong elastic band from it. Tie a yoghurt cup to this and mark where its top touches the board. Put packets that show their weight in ounces in turn into the cup. Mark on your scale where the cup touches the board giving the weight that stretched it there. This is called "calibrating." Put small objects into the cup and read the weight.

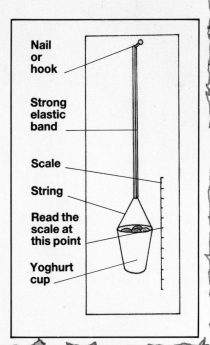

Nail or hook

Strong elastic band

Scale

String

Read the scale at this point

Yoghurt cup

LAUNCH A PLANE

Energy stored in a stretched elastic band can be used to launch a model glider. Ask an adult to cut a notch into a tongue depressor with a sharp knife and shape the stick to make the fuselage, or body, of the plane. Cut a large triangle shape from a postcard and glue it firmly onto the stick. Add a tail fin and squeeze a small blob of clay onto the nose of the plane to balance it.

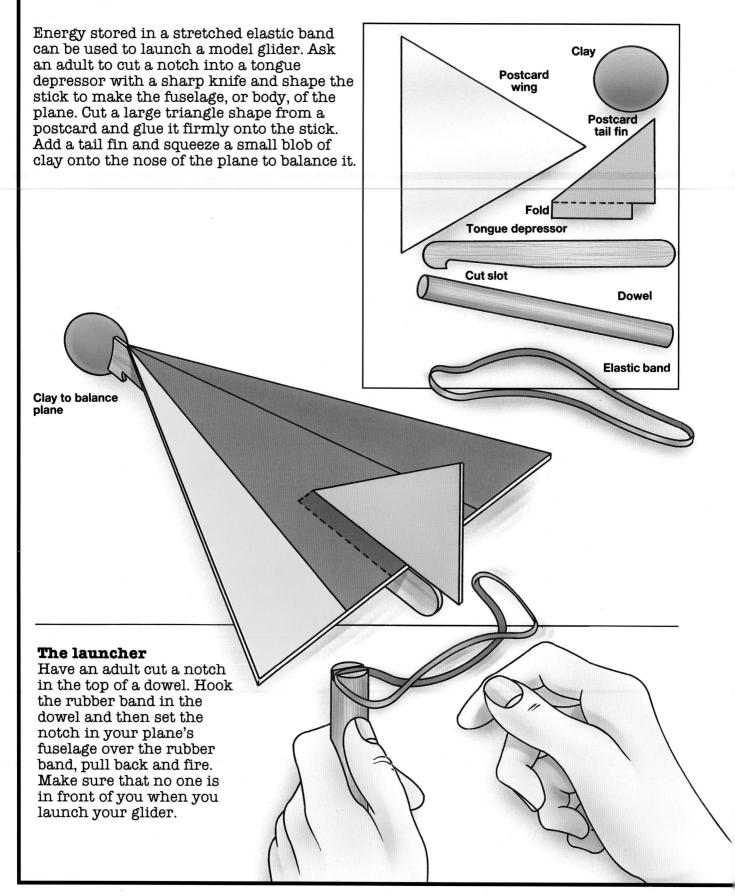

Clay

Postcard wing

Postcard tail fin

Fold

Tongue depressor

Cut slot

Dowel

Elastic band

Clay to balance plane

The launcher
Have an adult cut a notch in the top of a dowel. Hook the rubber band in the dowel and then set the notch in your plane's fuselage over the rubber band, pull back and fire. Make sure that no one is in front of you when you launch your glider.

FAIRGROUND CAROUSEL

This model carousel makes use of the energy stored in a twisted rubber band to power it.

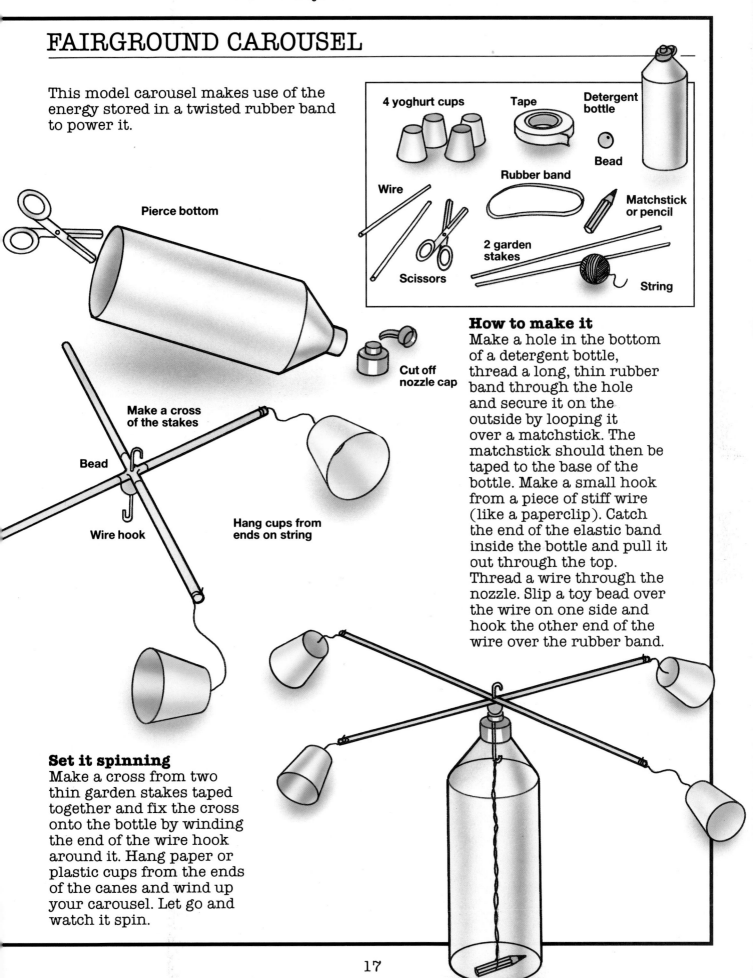

4 yoghurt cups

Tape

Detergent bottle

Bead

Wire

Rubber band

Matchstick or pencil

2 garden stakes

Scissors

String

Pierce bottom

Cut off nozzle cap

Make a cross of the stakes

Bead

Wire hook

Hang cups from ends on string

How to make it

Make a hole in the bottom of a detergent bottle, thread a long, thin rubber band through the hole and secure it on the outside by looping it over a matchstick. The matchstick should then be taped to the base of the bottle. Make a small hook from a piece of stiff wire (like a paperclip). Catch the end of the elastic band inside the bottle and pull it out through the top. Thread a wire through the nozzle. Slip a toy bead over the wire on one side and hook the other end of the wire over the rubber band.

Set it spinning

Make a cross from two thin garden stakes taped together and fix the cross onto the bottle by winding the end of the wire hook around it. Hang paper or plastic cups from the ends of the canes and wind up your carousel. Let go and watch it spin.

17

FLIGHT

Since people first walked on Earth they have dreamed of flying like the birds. Many brave inventors tried — and failed! Eventually, short flights were made in balloons and gliders, but it wasn't until the turn of the twentieth century that the first successful powered flight was made. Progress has been amazingly fast and today hundreds of travellers can fly thousands of miles in a few hours, in planes weighing hundreds of tons.

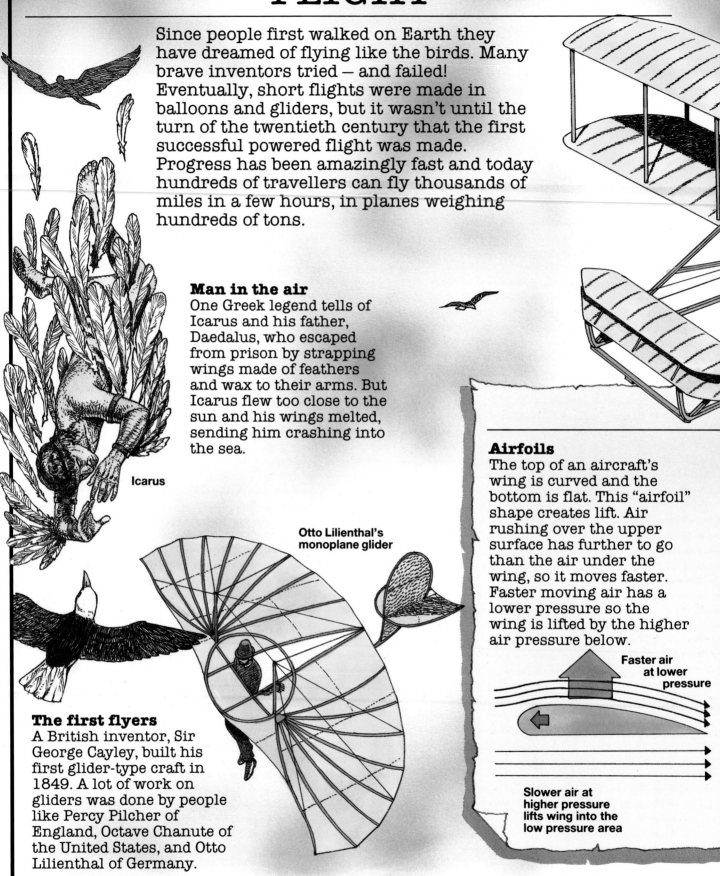

Man in the air
One Greek legend tells of Icarus and his father, Daedalus, who escaped from prison by strapping wings made of feathers and wax to their arms. But Icarus flew too close to the sun and his wings melted, sending him crashing into the sea.

Icarus

Otto Lilienthal's monoplane glider

The first flyers
A British inventor, Sir George Cayley, built his first glider-type craft in 1849. A lot of work on gliders was done by people like Percy Pilcher of England, Octave Chanute of the United States, and Otto Lilienthal of Germany.

Airfoils
The top of an aircraft's wing is curved and the bottom is flat. This "airfoil" shape creates lift. Air rushing over the upper surface has further to go than the air under the wing, so it moves faster. Faster moving air has a lower pressure so the wing is lifted by the higher air pressure below.

Faster air at lower pressure

Slower air at higher pressure lifts wing into the low pressure area

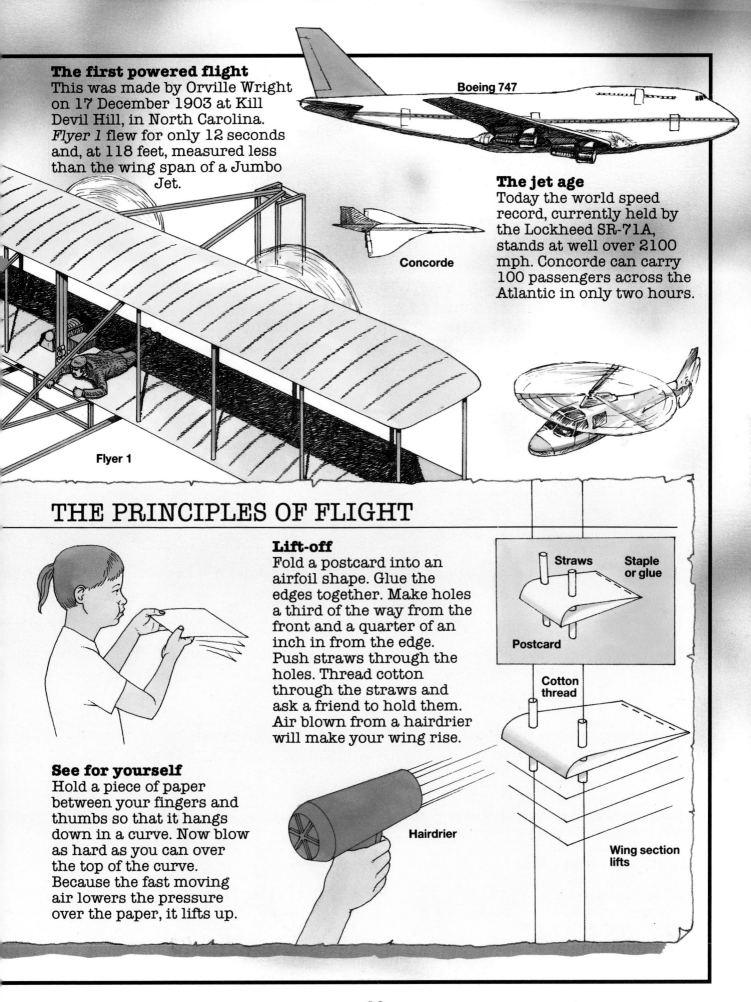

The first powered flight

This was made by Orville Wright on 17 December 1903 at Kill Devil Hill, in North Carolina. *Flyer 1* flew for only 12 seconds and, at 118 feet, measured less than the wing span of a Jumbo Jet.

Boeing 747

Concorde

The jet age

Today the world speed record, currently held by the Lockheed SR-71A, stands at well over 2100 mph. Concorde can carry 100 passengers across the Atlantic in only two hours.

Flyer 1

THE PRINCIPLES OF FLIGHT

Lift-off

Fold a postcard into an airfoil shape. Glue the edges together. Make holes a third of the way from the front and a quarter of an inch in from the edge. Push straws through the holes. Thread cotton through the straws and ask a friend to hold them. Air blown from a hairdrier will make your wing rise.

Straws **Staple or glue**

Postcard

Cotton thread

Hairdrier

Wing section lifts

See for yourself

Hold a piece of paper between your fingers and thumbs so that it hangs down in a curve. Now blow as hard as you can over the top of the curve. Because the fast moving air lowers the pressure over the paper, it lifts up.

BUILD YOUR OWN GLIDER

Sea birds like gulls soar easily into the sky, carried by the air currents around the cliffs where they live. Like gulls, gliders have long, slender wings and use upward moving air (called thermals), or air rising over hillsides, to carry them high into the sky without needing an engine.

How to make it
Glue the cardboard wing to the fuselage a third of the way from its end. Glue the tail to the fuselage and add the fin. Use a ball of clay on the nose of your glider as a balance weight.

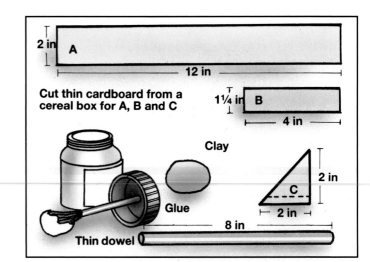

A — 2 in — 12 in

Cut thin cardboard from a cereal box for A, B and C

B — 1¼ in — 4 in

Clay

C — 2 in — 2 in

Glue

Thin dowel — 8 in

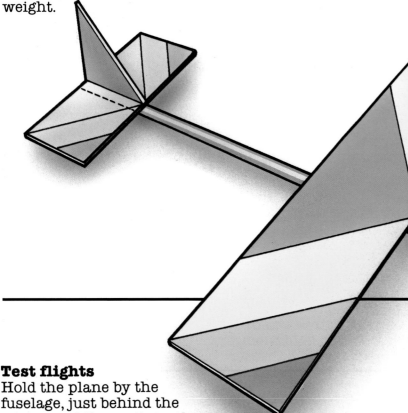

Balance weight

Test flights
Hold the plane by the fuselage, just behind the wings, and throw it gently forward. If the plane dives steeply, the balance weight is too heavy, so remove some of the clay and try again. If the plane climbs sharply, then stalls into a dive, add more weight to the nose. Change the balance weight until your plane glides in a gentle dive for a perfect flight.

Getting control

Building a plane that flies straight and level is only part of a designer's problem. A pilot needs to be able to steer the aircraft, to make it turn, climb or dive. This is done using flying controls.

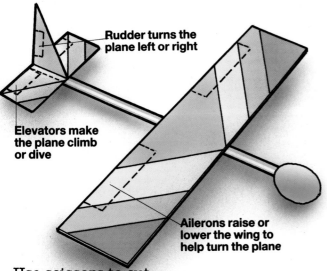

Rudder turns the plane left or right

Elevators make the plane climb or dive

Ailerons raise or lower the wing to help turn the plane

Use scissors to cut ailerons, elevators and rudder controls in your glider. Test-fly your plane with different control settings, e.g. try bending the left aileron down, the right aileron up and bend the rudder to the right. Your glider should bank and turn to the right.

MAKE A PAPER PLANE

Try different sizes of wing by folding down more or less of the wing flaps. Also experiment with different balance weights on the nose by adding paperclips to the front.

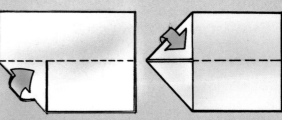

Fold in the corners

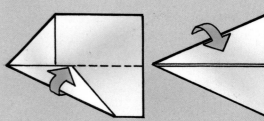

Fold over the wings

Fold the plane in half

Fold down the wings

Plane dives – balance weight too heavy

Plane climbs steeply and stalls – balance weight too light

Gentle glide – perfect flight

MAKE A KITE

Kites have been made and flown for hundreds and perhaps thousands of years in countries like China. Some kites are huge, take months to make and need teams of people to fly them, but you can make your own simple kite very easily. Cut a large kite shape from a plastic bag. Tie two garden stakes together with thin string. Lay the crossed sticks onto the plastic kite and tie the points of the kite to the ends of the sticks. Tape ribbon to the bottom of the kite. This is its tail and helps to balance it. Tie strings to each end of the sticks and join them to a ball of string underneath the kite.

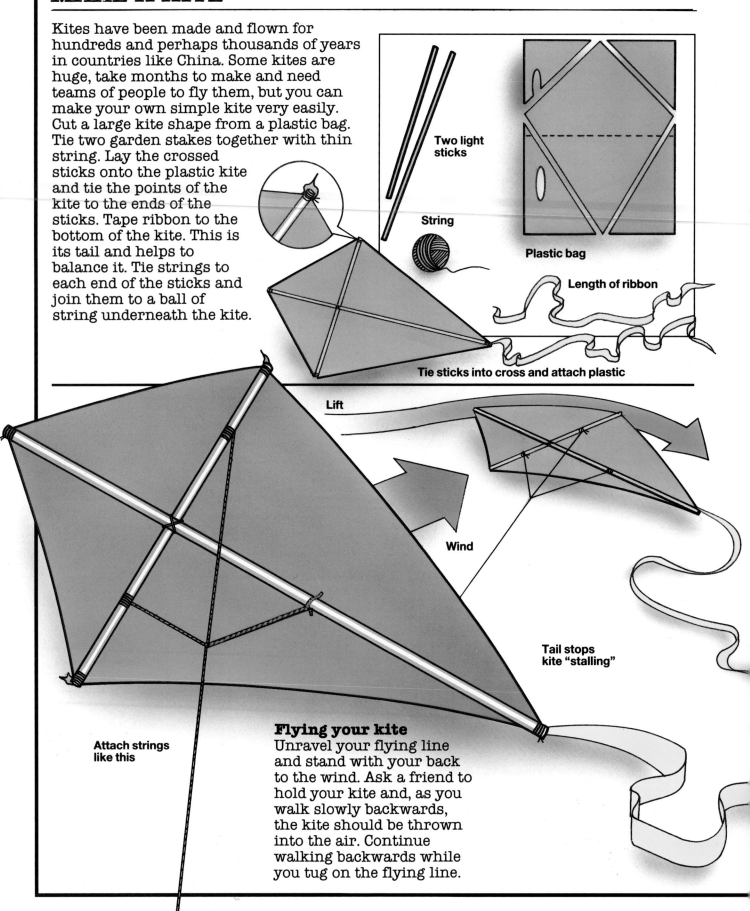

Two light sticks

String

Plastic bag

Length of ribbon

Tie sticks into cross and attach plastic

Lift

Wind

Tail stops kite "stalling"

Attach strings like this

Flying your kite
Unravel your flying line and stand with your back to the wind. Ask a friend to hold your kite and, as you walk slowly backwards, the kite should be thrown into the air. Continue walking backwards while you tug on the flying line.

MAKE A HELICOPTER

You can make your own helicopter spinner quite easily. Take a cocktail stick or drinking straw. Cut helicopter blades from some cardboard — you must cut out the whole thing from one piece — and secure them to the stem with a thumbtack or blob of clay. The blades need to be twisted like a propeller to catch the air.

How to spin it
Hold the stem between your hands and set it spinning by rubbing one hand against the other. If your spinner doesn't fly upwards, spin it the other way. Now the blades will catch the air and your helicopter spinner will fly away.

Stiff card

Clay

Plastic straw

Spin between hands

MAKE A HOVERCRAFT

A hovercraft is a flying machine that is carried on a cushion of air. A powerful engine drives a fan that blows air downwards. The air is trapped under a rubber "skirt" that runs round the edge of the hovercraft, lifting it clear of the ground. Engine power is also used to drive a propeller to push it along at high speed.

To make a simple model hovercraft, cut a circle out of the center of a shoebox lid. Lay the lid on a smooth, flat surface. Ask an adult to help you blow air through the hole with a hairdrier. The lid will rise on its cushion of air, and a slight nudge will get it skimming along.

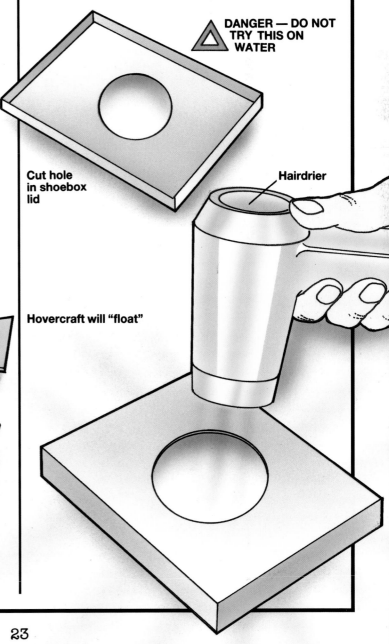

DANGER — DO NOT TRY THIS ON WATER

Cut hole in shoebox lid

Hairdrier

Hovercraft will "float"

NATURAL ENERGY  (WIND AND WATER)

You cannot see it, feel it, smell it or taste it, but it is all around you — air. The only time we really notice air is when it moves. Wind rustles the leaves on trees and ruffles your hair, but it can also blow so hard that it can uproot trees and tear down buildings. The power of moving air has been used for centuries as a natural source of energy to power machinery and drive sailing ships.

Windmills
Windmills use the wind to grind corn or pump water or work machines. The wind turns large sails or sweeps which are connected to drive shafts by large wooden cogs.

Sail power
Since the time of the ancient Egyptians, Greeks and Romans, wind power has been used to drive ships. Wind power, unlike fuel for engine-driven craft, is free and unlimited.

PRINCIPLES OF NATURAL ENERGY

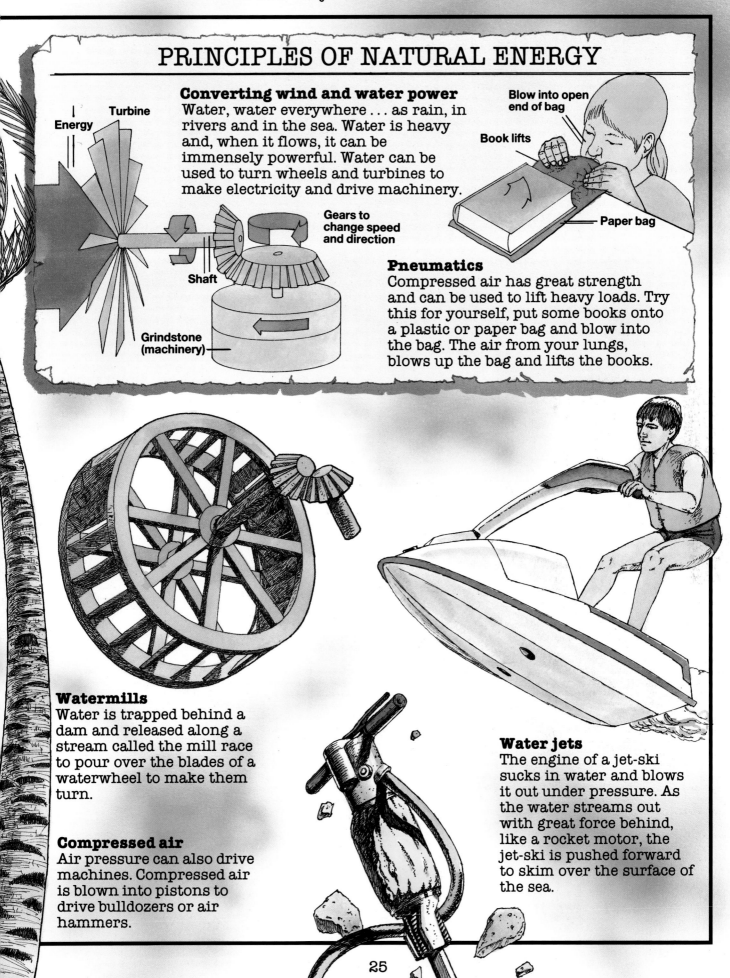

Converting wind and water power

Water, water everywhere . . . as rain, in rivers and in the sea. Water is heavy and, when it flows, it can be immensely powerful. Water can be used to turn wheels and turbines to make electricity and drive machinery.

Energy

Turbine

Gears to change speed and direction

Shaft

Grindstone (machinery)

Blow into open end of bag

Book lifts

Paper bag

Pneumatics

Compressed air has great strength and can be used to lift heavy loads. Try this for yourself, put some books onto a plastic or paper bag and blow into the bag. The air from your lungs, blows up the bag and lifts the books.

Watermills

Water is trapped behind a dam and released along a stream called the mill race to pour over the blades of a waterwheel to make them turn.

Compressed air

Air pressure can also drive machines. Compressed air is blown into pistons to drive bulldozers or air hammers.

Water jets

The engine of a jet-ski sucks in water and blows it out under pressure. As the water streams out with great force behind, like a rocket motor, the jet-ski is pushed forward to skim over the surface of the sea.

MAKE A WINDMILL

When people think of windmills they think of millers grinding corn. But many windmills were built to pump water, particularly in Holland and the English fens, in order to drain the land.

How to make it

Cut and fold some paper to make the sails. Push stiff wire through the center of the sails and use the jet from a detergent bottle or a toy bead as a spacer between the box and the sails, with a plastic straw as a spacer inside the box. Finally add a hammer cut from stiff cardboard and a cam, also cut from cardboard and taped to the end of the wire that is attached to the sails.

How it works

As the wind blows, it catches the sails and makes them spin. Because the sails and the cam are joined by the wire, the cam turns with the sails. Each time the raised section of the cam swings round and hits the hammer, the hammer is raised and then falls back again. This mechanism is called a trip hammer.

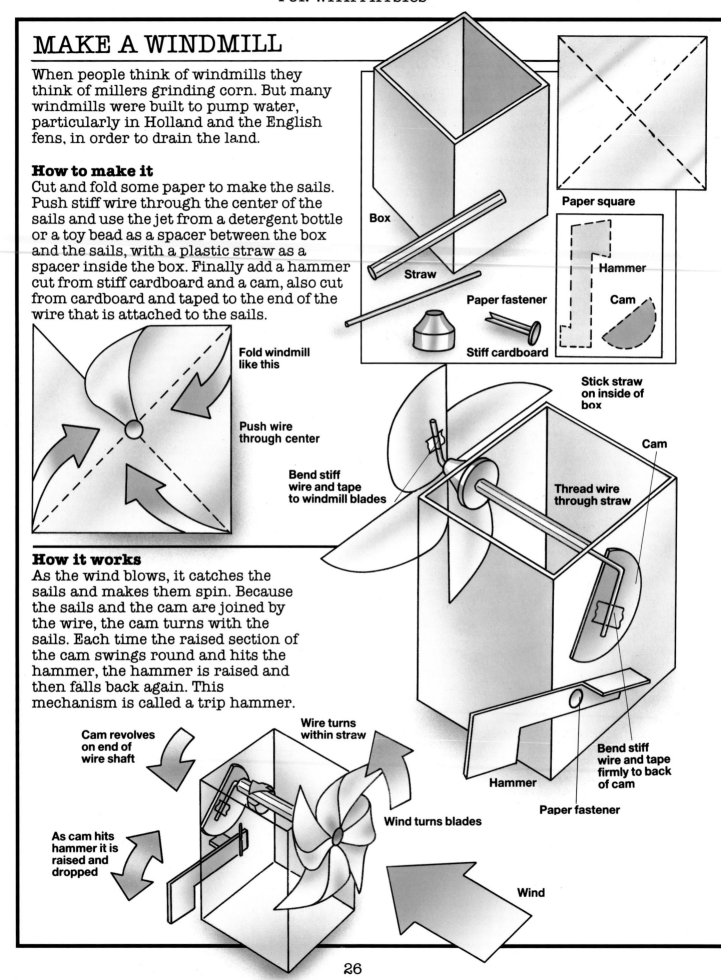

Box

Straw

Paper square

Hammer

Cam

Paper fastener

Stiff cardboard

Fold windmill like this

Push wire through center

Bend stiff wire and tape to windmill blades

Stick straw on inside of box

Cam

Thread wire through straw

Bend stiff wire and tape firmly to back of cam

Hammer

Paper fastener

Cam revolves on end of wire shaft

Wire turns within straw

As cam hits hammer it is raised and dropped

Wind turns blades

Wind

WATERWHEEL LIFTING GEAR

How to make it

Cut four large flaps into the side of a detergent bottle and bend the flaps outwards. Push a piece of stiff wire or a knitting needle through the center of the base of the bottle and out through the nozzle. Make sure that it can spin freely. Now tape a length of string to one end of the bottle with a weight tied to the end. To operate your waterwheel, hold the blades under a steam of fast running water.

Detergent bottle

String

Clay weight

Long knitting needle or stiff wire

Scissors

Cut and fold out four flaps

Cut off nozzle cap

Push needle through nozzle and out of the bottom

Tape string to side of bottle

Running water

How it works

The force of the water hitting the blades causes the bottle to spin round. As it turns it winds up the string and lifts the weight. Can you get it to go down again?

Bottle revolves

Weight is raised

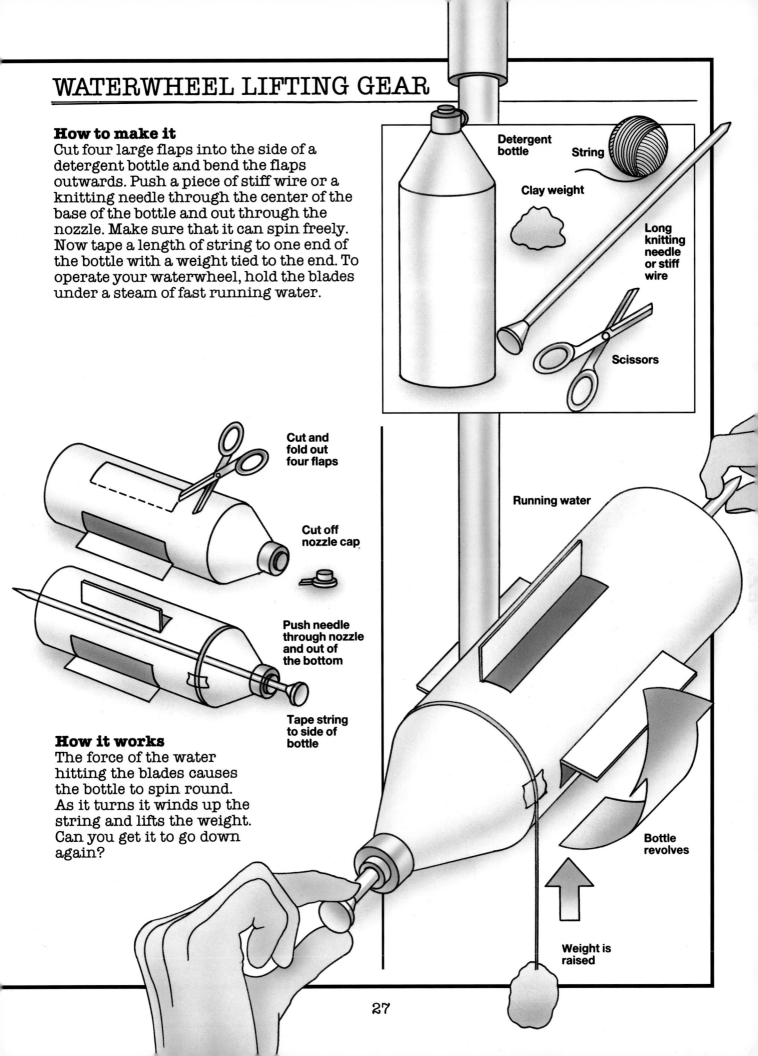

AIR-POWERED ROCKET

When you blow up a balloon you are squeezing or compressing air inside an elastic skin. If you let the balloon go, the air is forced out through the neck with a rush that pushes the balloon in the opposite direction, making it fly in an erratic and amusing way.

The power of the air can be released more slowly by taping a short piece of tube like a jumbo drinking straw into the neck of the balloon. Put a long piece of cotton or thin twine through a straw, stretch the cotton across a room, and tape your balloon to the straw. Blow up the balloon and, when it is released, the escaping air will push the balloon along the string.

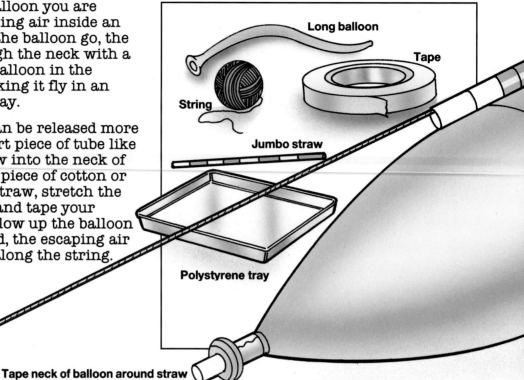

String
Long balloon
Tape
Jumbo straw
Polystyrene tray
Tape neck of balloon around straw

PLASTIC BOTTLE BOAT

With a few simple objects you can make a Viking-style sailing ship with a large square sail to catch the wind. The addition of a keel stabilizes the boat.

Use a pair of scissors to cut a plastic bottle into an interesting boat shape. Use a thin stake for the mast and tape a sail to it, cut from a plastic bag or a sheet of paper. The mast can be set upright inside the boat using a blob of clay.

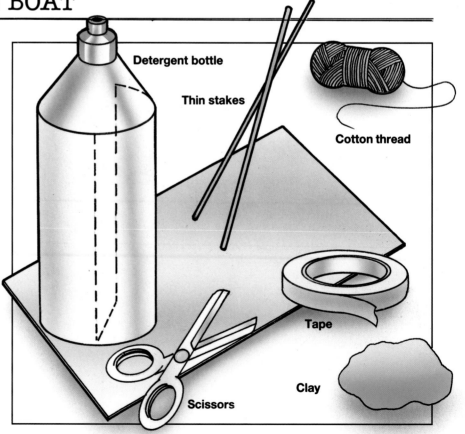

Detergent bottle
Thin stakes
Cotton thread
Tape
Scissors
Clay

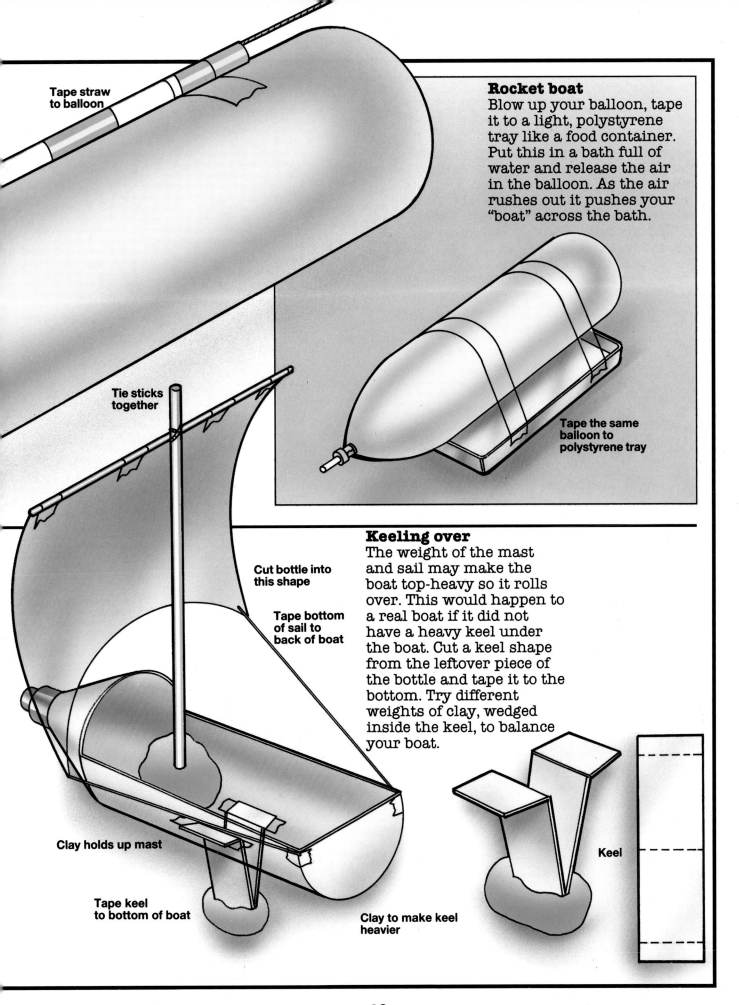

Tape straw to balloon

Tie sticks together

Rocket boat

Blow up your balloon, tape it to a light, polystyrene tray like a food container. Put this in a bath full of water and release the air in the balloon. As the air rushes out it pushes your "boat" across the bath.

Tape the same balloon to polystyrene tray

Cut bottle into this shape

Tape bottom of sail to back of boat

Keeling over

The weight of the mast and sail may make the boat top-heavy so it rolls over. This would happen to a real boat if it did not have a heavy keel under the boat. Cut a keel shape from the leftover piece of the bottle and tape it to the bottom. Try different weights of clay, wedged inside the keel, to balance your boat.

Clay holds up mast

Tape keel to bottom of boat

Clay to make keel heavier

Keel

29

WIND SPEED AND DIRECTION INDICATOR

How to make an anemometer

You can make your own weather instruments very easily. An anemometer measures wind speed. Push a rounded pencil or a piece of dowel into the end of a large garden cane that has been knocked into the ground in a suitable spot. Make a hole in the base of a plastic yoghurt cup that is just large enough to slip over the dowel. Split the end of some plastic drinking straws and tape or glue them to the sides of the cup. Glue light paper or polystyrene cups to the ends of the straws. Add strengthening pieces to hold the cups rigidly in place.

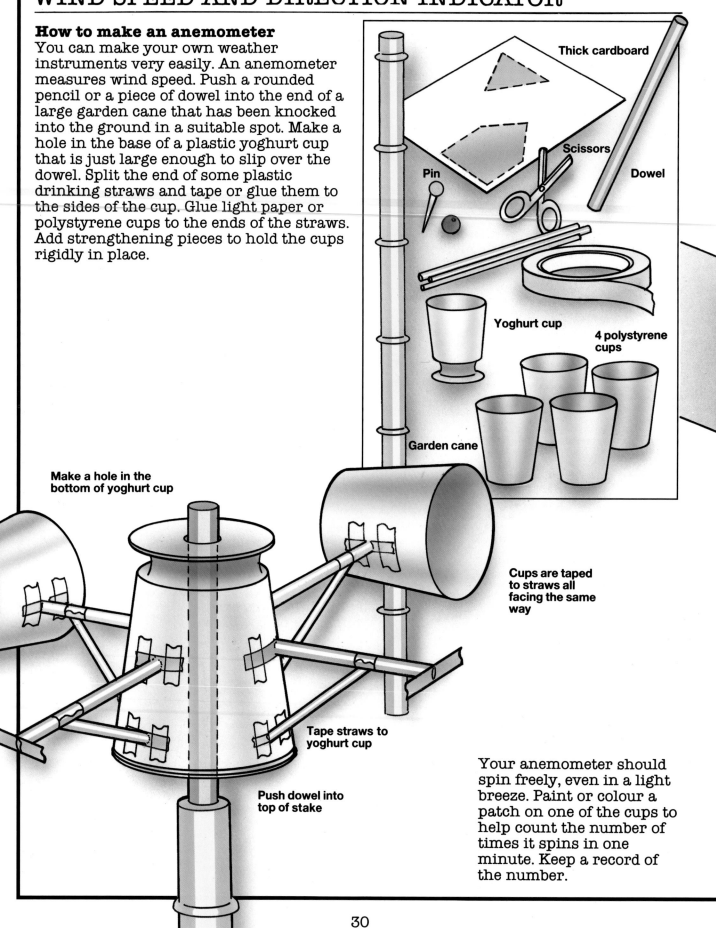

Thick cardboard

Scissors

Pin

Dowel

Yoghurt cup

4 polystyrene cups

Garden cane

Make a hole in the bottom of yoghurt cup

Cups are taped to straws all facing the same way

Tape straws to yoghurt cup

Push dowel into top of stake

Your anemometer should spin freely, even in a light breeze. Paint or colour a patch on one of the cups to help count the number of times it spins in one minute. Keep a record of the number.

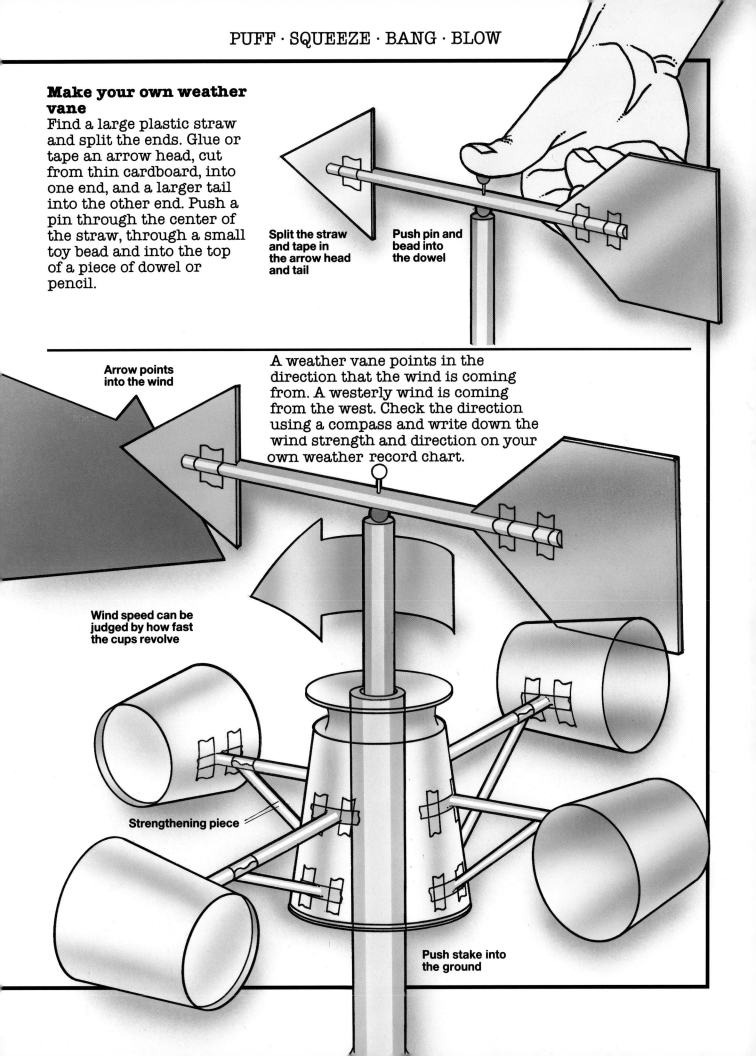

Make your own weather vane

Find a large plastic straw and split the ends. Glue or tape an arrow head, cut from thin cardboard, into one end, and a larger tail into the other end. Push a pin through the center of the straw, through a small toy bead and into the top of a piece of dowel or pencil.

Split the straw and tape in the arrow head and tail

Push pin and bead into the dowel

Arrow points into the wind

A weather vane points in the direction that the wind is coming from. A westerly wind is coming from the west. Check the direction using a compass and write down the wind strength and direction on your own weather record chart.

Wind speed can be judged by how fast the cups revolve

Strengthening piece

Push stake into the ground

SOUND OF MUSIC

Sound is made when something is made to shake or vibrate. Place your fingertips on your throat and speak. You can feel the vibrations of your own voice. Quiet sounds produce tiny vibrations; very noisy things shake the air violently. All sounds reach us through the air. The vibrating object shakes the air and the vibrations spread through the air like ripples on a pond. Sound can also be carried through solid objects and liquids. Where there is no air and nothing to transmit sound, such as in the vacuum of deep space, you would hear nothing at all.

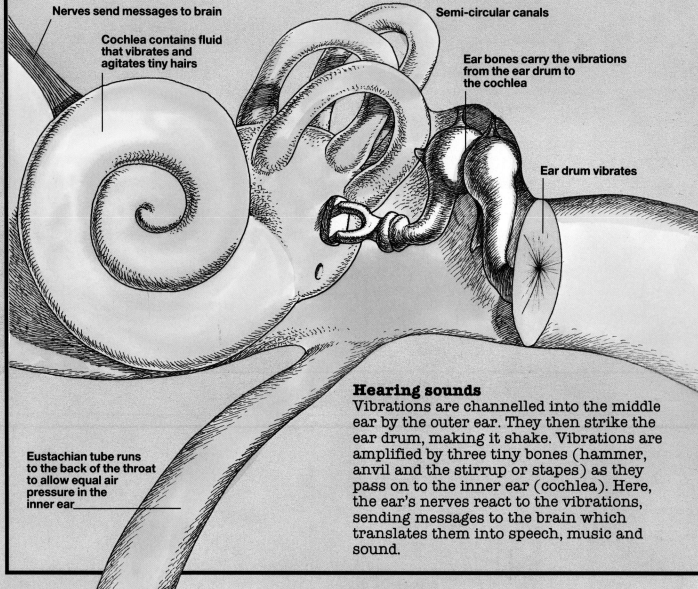

Nerves send messages to brain

Cochlea contains fluid that vibrates and agitates tiny hairs

Semi-circular canals

Ear bones carry the vibrations from the ear drum to the cochlea

Ear drum vibrates

Eustachian tube runs to the back of the throat to allow equal air pressure in the inner ear

Hearing sounds

Vibrations are channelled into the middle ear by the outer ear. They then strike the ear drum, making it shake. Vibrations are amplified by three tiny bones (hammer, anvil and the stirrup or stapes) as they pass on to the inner ear (cochlea). Here, the ear's nerves react to the vibrations, sending messages to the brain which translates them into speech, music and sound.

PRINCIPLES OF SOUND

Good vibrations

Blow up a balloon and hold it against your ear while a friend speaks to you through the balloon. Press your lips gently against the balloon and speak back. Not only can you hear the vibrations through the balloon, but you can feel your own voice through your lips as the balloon's skin vibrates against them.

See for yourself

Lay a radio or cassette player on its back and sprinkle scraps of tissue onto the speaker. Turn up the volume and the tissue will dance, driven by the vibrations of the music.

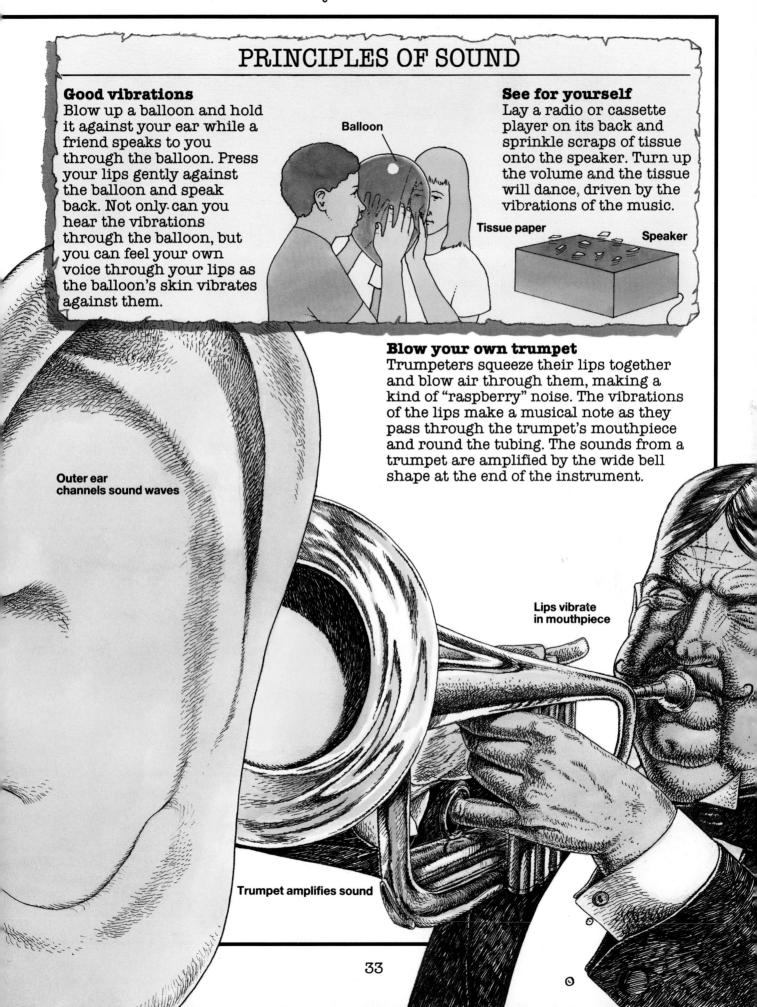

Balloon

Tissue paper

Speaker

Blow your own trumpet

Trumpeters squeeze their lips together and blow air through them, making a kind of "raspberry" noise. The vibrations of the lips make a musical note as they pass through the trumpet's mouthpiece and round the tubing. The sounds from a trumpet are amplified by the wide bell shape at the end of the instrument.

Outer ear channels sound waves

Lips vibrate in mouthpiece

Trumpet amplifies sound

YOGHURT-CUP TELEPHONE

Use a pencil to make a hole in the bottom of two empty yoghurt cups. Thread a long piece of string through each hole and tie knots in them to stop the string from pulling back out. Give one pot to a friend, then stand far apart so that the string is stretched tight. Speak loudly and clearly into your pot while your friend listens into the other one. Now you listen as your friend speaks.

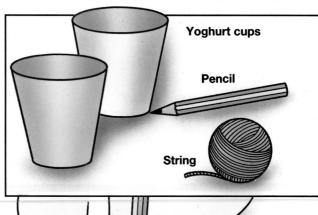

Yoghurt cups

Pencil

String

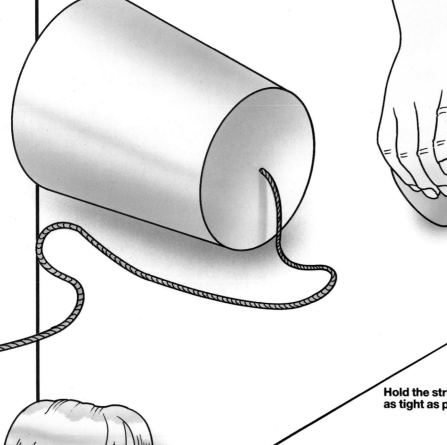

Speaker

Hold the string as tight as possible

Listener

You can hear voices clearly through your telephone because the vibrations of a person's voice make the pot shake. These vibrations are passed along the string to the other pot, which also begins to vibrate. This shakes the air in the second pot and you recognize the vibrations as your friend's voice.

YOGHURT-CUP RECORD PLAYER

Push a pin through the bottom of a plastic cup with a pair of pliers. Hold the pin in the grooves of an old, unwanted record that is spinning on the deck of a record player. Listen closely to the pot and you will hear the recorded music clearly. The record's grooves have tiny bumps along their sides which hit the tip of the pin, making it vibrate. The pot amplifies these tiny sounds, which you hear as music.

STETHOSCOPE

Push two plastic funnels into the ends of a length of plastic tubing. Ask a friend to hold one end over their heart or a ticking watch or some other tiny sound while you listen carefully in the other funnel. The tiny vibrations are collected by the funnel and directed along the tube towards your ear. In a very quiet room, you will be able to hear the sound clearly.

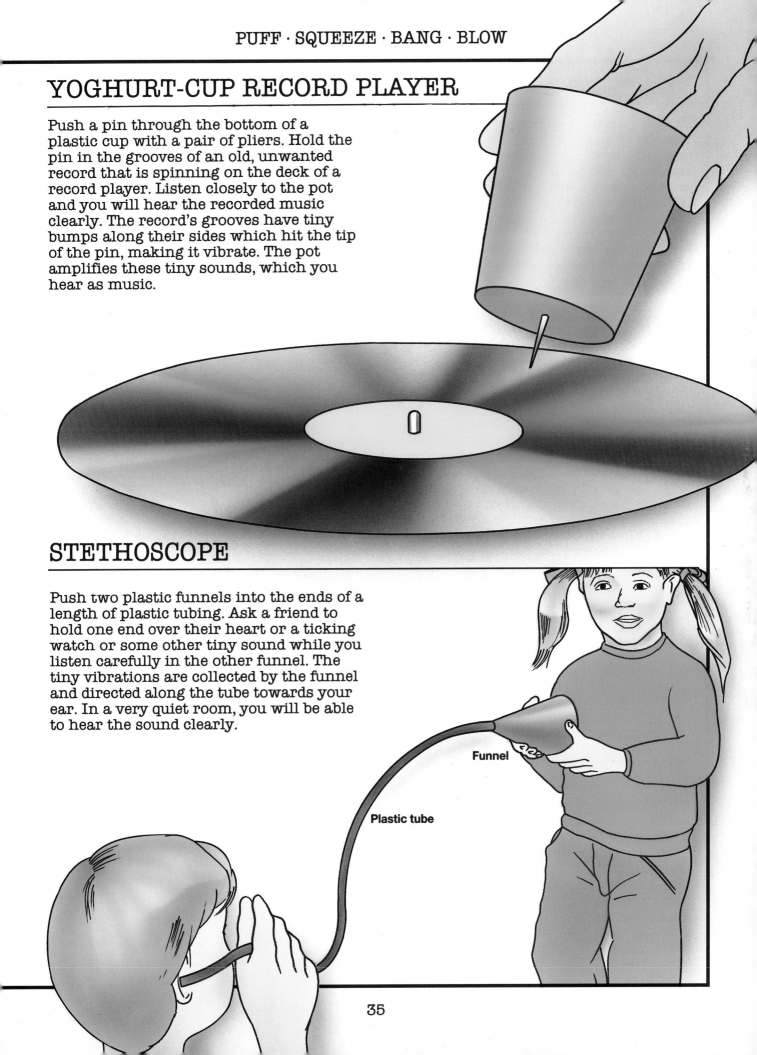

Funnel

Plastic tube

35

SHAKERS

Choose hard plastic pots of various shapes and sizes and put some rice, pasta, small stones and dried peas or beans inside. Experiment with the various sounds that you can get by shaking different things inside your pots. When you have found the best combination of pot and filling, find a length of wood that is the right size to fit tightly into the neck of the pot as a handle. Now you can really play your maracas.

Decorate bottles

Small plastic bottles

Fill with rice, pasta or dried peas

Sticks

DRUMS

Make your own drums by finding some large containers such as waste paper bins or large plastic or tin boxes. Cut circles of plastic from strong trash bags, 4 inches larger all round than the size of the container. Ask a friend to help you stretch the skin tightly over the open end of your drum while you tie or tape it firmly. The tighter the skin, the better the sound. Use sticks as drum beaters.

Plastic bin liner tied tightly over container

Sticks

You can use a waste paper basket, mixing bowl or food container

BOTTLE BAND

Tap the sides of some glass bottles gently with a wooden spoon and listen to the ringing notes that they play. Add different amounts of water to your bottles to see how the notes change. With care you can "tune" your bottles and jars and play simple melodies.

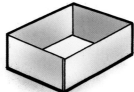

Fill bottles with different amounts of water

2

CLICK FLASH BUZZ WHIRR

THE SCIENCE AND TECHNOLOGY OF ELECTRICITY AND MAGNETISM

MAGNETISM

Magnets have many uses and come in all shapes and sizes. They are not just used in compasses, they are also used inside televisions, stereos, radios, telephones and other everyday objects. Catches on some cupboard doors are magnetic, and every electric motor in toys, machines and vehicles has a strong magnet inside it.

Magnets for sound

Sounds are vibrations. The speakers in radios, TVs, telephones and stereos contain strong magnets that cause a disc called a diaphragm to shake or vibrate, sending out sound waves that we hear as speech or music.

The lodestone

The heavy rock that attracts iron is special rock called magnetite or lodestone, which means the "leading stone." Some sailors hung pieces of it from cord and used it like a compass to help them navigate.

Finding your way

A compass has a magnetized pointer, or needle, that turns inside a protective container marked with the points of the compass. The needle swings until it points towards magnetic north.

Sealing doors

Magnets are often used on the doors and lids of fridges and freezers to make them airtight. The rubber seal on the door has a magnetized strip inside it that is attracted to the steel of the main part of the fridge, pulling the door tightly shut.

PRINCIPLES OF MAGNETISM

Poles of a magnet

All magnets have two poles, the north-seeking pole and the south-seeking pole. A magnet will attract another magnet only if two opposite poles are brought together. Two similar poles repel or push each other away.

North
South

South

South

Repelling

Attracting

South

North

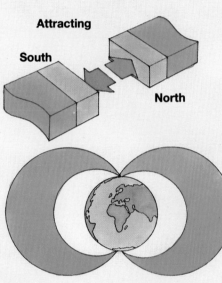

Fields around Earth

Our planet is like a gigantic magnet. It has a molten core of iron and a powerful magnetic field surrounds it. Force lines run roughly from the north pole to the south pole, but currents in the iron core cause slight changes in the position of magnetic north and south.

Finding magnetic fields

Lay a sheet of white paper over a magnet and sprinkle some iron filings over the top. Gently tap the sheet with your finger. You will see the iron filings form into curves as they follow the force lines that make up the magnetic field around a magnet.

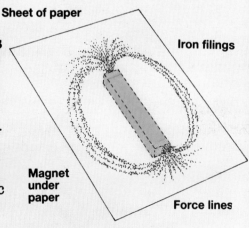

Sheet of paper

Iron filings

Magnet under paper

Force lines

Cotton thread

**Magnet swings freely
until it lies
in a north-south
direction**

**Paper or cardboard
sling**

S

N

MAKE A COMPASS

Stroke a large, steel needle with the south pole of a strong magnet. Hold the needle firmly with your finger tips as you run the magnet along the needle from the hole to the tip. Repeat this about 10 times, moving the magnet well away from the needle between strokes.

Magnetic materials, like a steel needle, are made up of tiny magnetic particles called domaines. These are usually lying jumbled up, but stroking with a magnet makes them line up in the same direction, turning the needle into a small magnet.

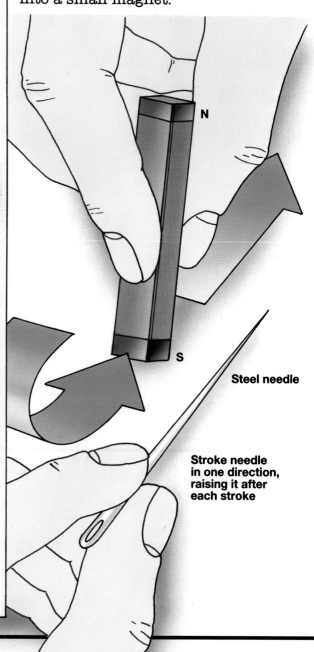

N

S

Steel needle

**Stroke needle
in one direction,
raising it after
each stroke**

SUSPENDED MAGNET

Cut a loop of stiff paper or thin cardboard to make a sling to hold your magnet. Hang the sling from a piece of cotton thread and let it swing freely. When the magnet stops swinging, note the direction in which it is pointing. Do this a few times, what do you discover? When a suspended magnet stops swinging it is always pointing in the north/south direction. The magnet's north pole should really be called the north-seeking pole because it will always swing to point towards Earth's magnetic north.

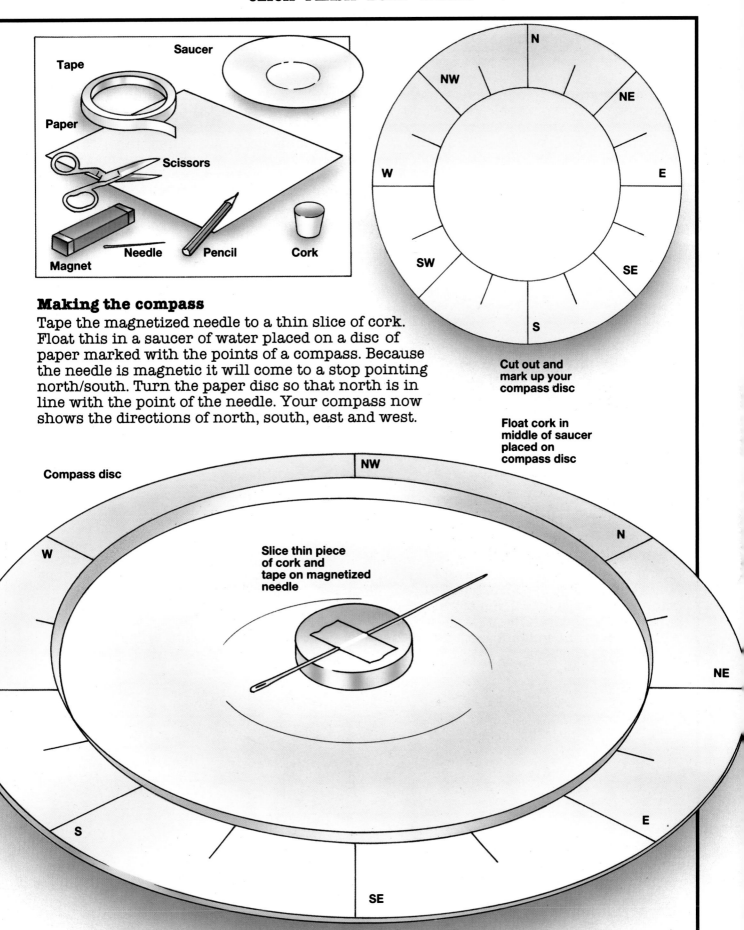

Making the compass

Tape the magnetized needle to a thin slice of cork. Float this in a saucer of water placed on a disc of paper marked with the points of a compass. Because the needle is magnetic it will come to a stop pointing north/south. Turn the paper disc so that north is in line with the point of the needle. Your compass now shows the directions of north, south, east and west.

Cut out and mark up your compass disc

Float cork in middle of saucer placed on compass disc

Slice thin piece of cork and tape on magnetized needle

Compass disc

MAGNETIC SORTING

Magnets are only attracted to other magnetic materials. To find out what these are, collect some objects like bottle tops, pins, keys, stones, rubber, paper, tin foil, things made from plastic and glass, coins, nails, nuts and bolts and so on. Use a magnet to sort your collection. Find all the things that seem to "stick" to your magnet. What are these made from?

Looking after magnets

Magnets are made by lining up the tiny magnetic domains in the metal in the same direction. Dropping a magnet will upset the particles and destroy the magnetism. When they are not being used, store them in pairs with the north pole of one against the south pole of the other. "Keepers" can be placed across the poles to help stop the magnetism from "leaking" away.

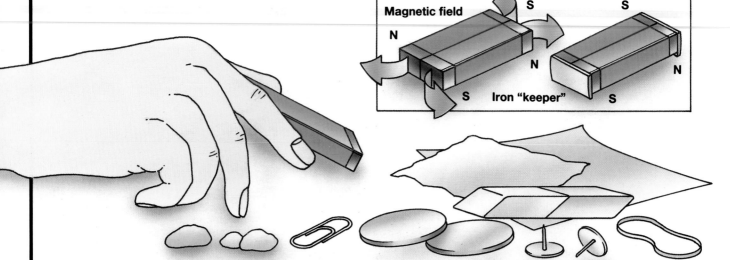

Magnetic field

Iron "keeper"

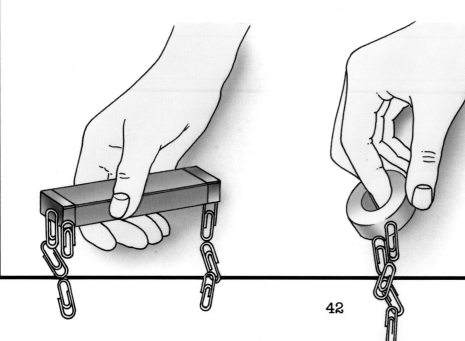

TESTING STRENGTH

Where is it strongest?

Which part of a magnet is the strongest? Is it in the middle or at the ends? Dip a magnet sideways into a pile of pins or paperclips and carefully lift it clear. To which part of the magnet are most of the pins or clips attracted?

Which one is strongest?

Compare the strength of different types of magnet by dipping each one in turn into a pile of paperclips. Carefully lift the magnet clear and count how many clips are left clinging to the magnet. A good scientist would always repeat the test a number of times to take the average. Will a magnet pick up the same number each time?

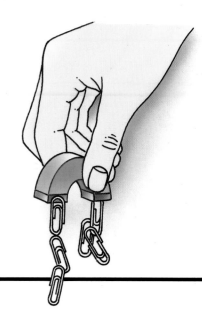

THE TUMBLER TRICK

Magnets can attract magnetic objects from some distance away. They can also work through water and glass. Tell your friends that you can get a paperclip out of a glass of water without getting it wet. Put the pole of a strong magnet to the glass and slide it slowly up the side. The paperclip, attracted by the magnet, will follow it up the glass and pop out at the top.

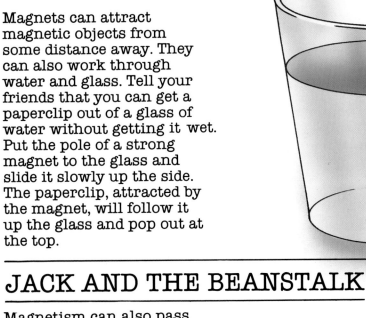

JACK AND THE BEANSTALK

Magnetism can also pass through things such as cloth, paper and cardboard. Cut out the shape of Jack from a piece of paper and tape a paperclip to his back. Draw the beanstalk onto a piece of thin cardboard and place Jack at the bottom. Hold a strong magnet behind the cardboard, close to where you have placed Jack. As you move your magnet slowly up the cardboard, Jack will seem to climb magically up the beanstalk.

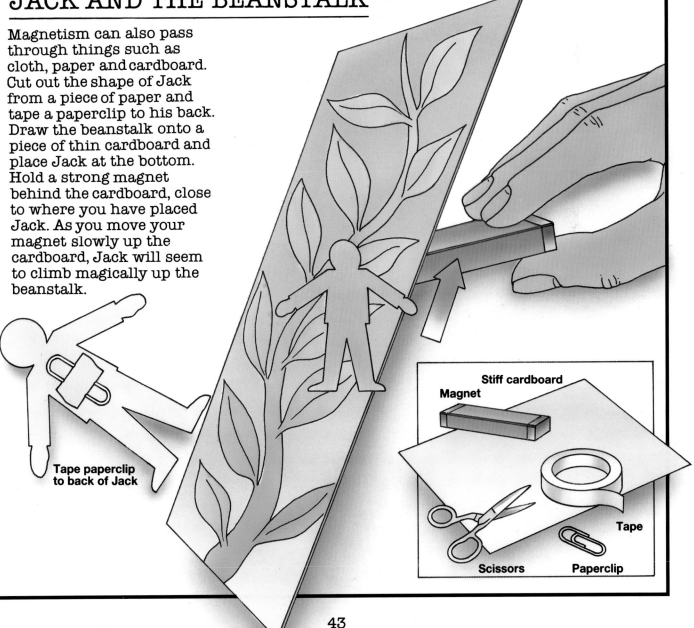

Tape paperclip to back of Jack

Stiff cardboard

Magnet

Scissors

Tape

Paperclip

FLOATING MAGNET

If you are lucky enough to have two strong magnets you can show how the poles can repel or push each other apart in a very dramatic way.

What to do

Lay one of the magnets on a piece of balsa wood or thick polystyrene packing material. Push cocktail sticks into the bottom of this, around the edge of the magnet. Use an eraser to help you, so that you don't prick your finger on the point of the sticks. Carefully lower the second magnet onto the first, making sure that the two north poles and the two south poles are lined up. When the magnet is released it will bob up and float in mid-air, held by the powerful repelling force between the poles.

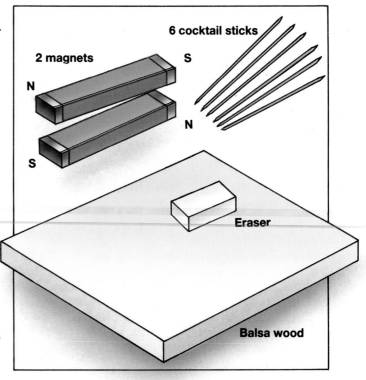

6 cocktail sticks

2 magnets

S

N

N

S

Eraser

Balsa wood

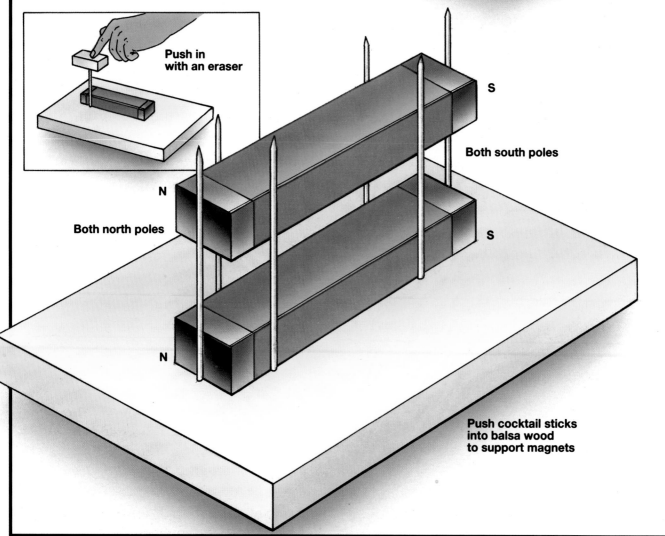

Push in
with an eraser

Both south poles

S

S

Both north poles

N

N

Push cocktail sticks
into balsa wood
to support magnets

FLYING PAPERCLIP

Paperclips or other small steel objects can be made to defy gravity using the power of magnetic attraction.

Tape a magnet to a wall or a shelf above a table top. Tie a length of cotton thread to a paperclip and fasten it to a piece of wood with a thumbtack. Make sure that the cotton thread is the right length so that when the paperclip is held up it will not quite reach the magnet, but remain floating mysteriously under it.

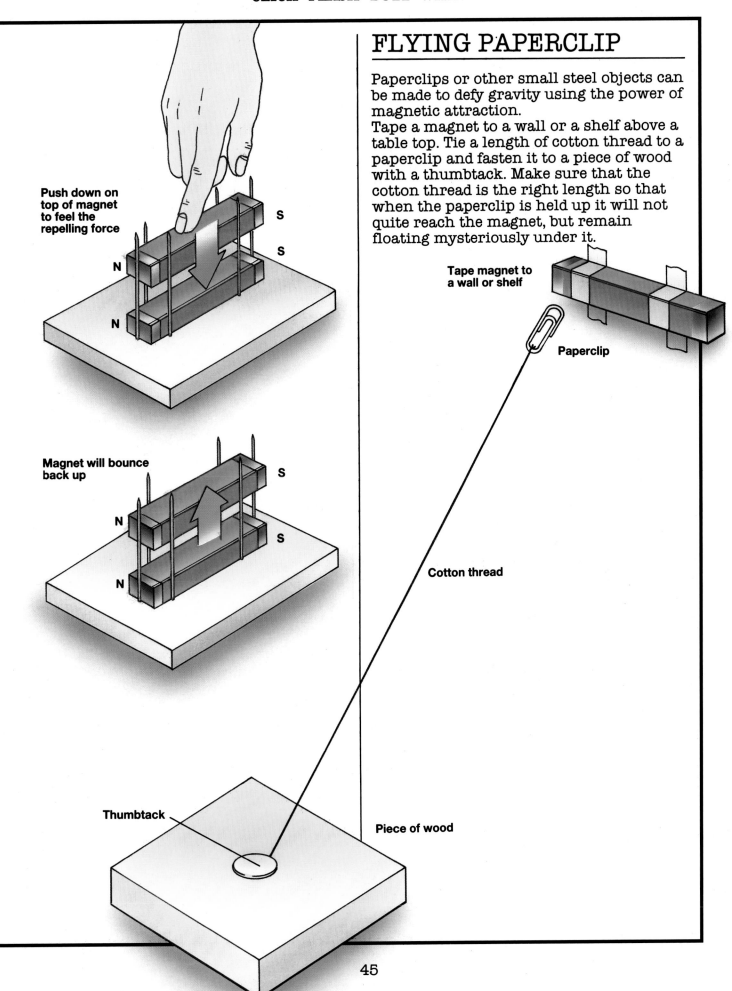

Push down on top of magnet to feel the repelling force

S

S

N

N

Magnet will bounce back up

S

N

S

N

Tape magnet to a wall or shelf

Paperclip

Cotton thread

Thumbtack

Piece of wood

45

FISHING GAME

This is an interesting variation of the popular magnetic game.

Use a large plastic bowl for the fish tank and fill it with water. Add some blue ink or food coloring to the water to hide the fish. Make your fish, sharks and an old boot by cutting shapes from an old plastic bag. Push paperclips onto the fish and the boot shapes. This makes them sink. The fish will be caught when the magnetic rod attracts the paperclips.

How to play

Take turns fishing in the water with your magnetic rod. If you catch a fish you score 5 points, a shark scores 10, but beware of the boot, this will lose you 20 points. The first to score 30 points wins the game! Remember to dry your magnet carefully after the game.

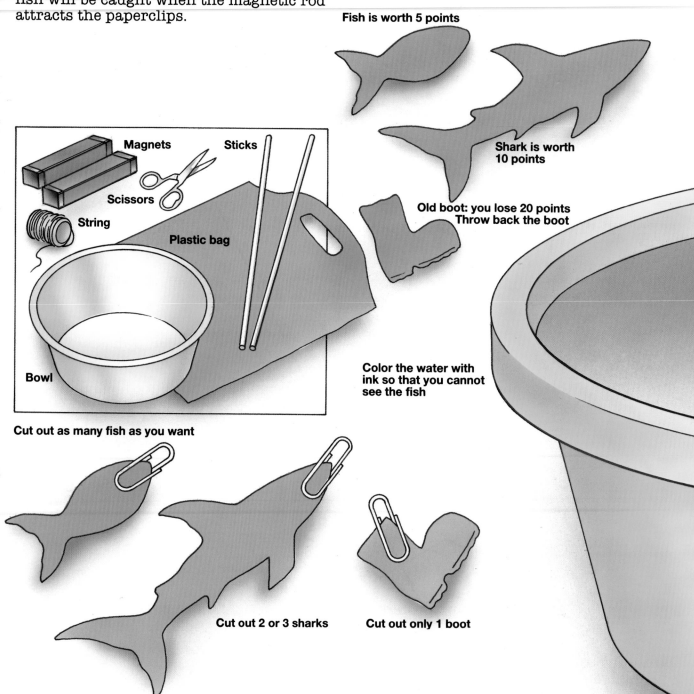

Fish is worth 5 points

Shark is worth 10 points

Old boot: you lose 20 points
Throw back the boot

Magnets **Sticks**

Scissors

String

Plastic bag

Bowl

Color the water with ink so that you cannot see the fish

Cut out as many fish as you want

Cut out 2 or 3 sharks

Cut out only 1 boot

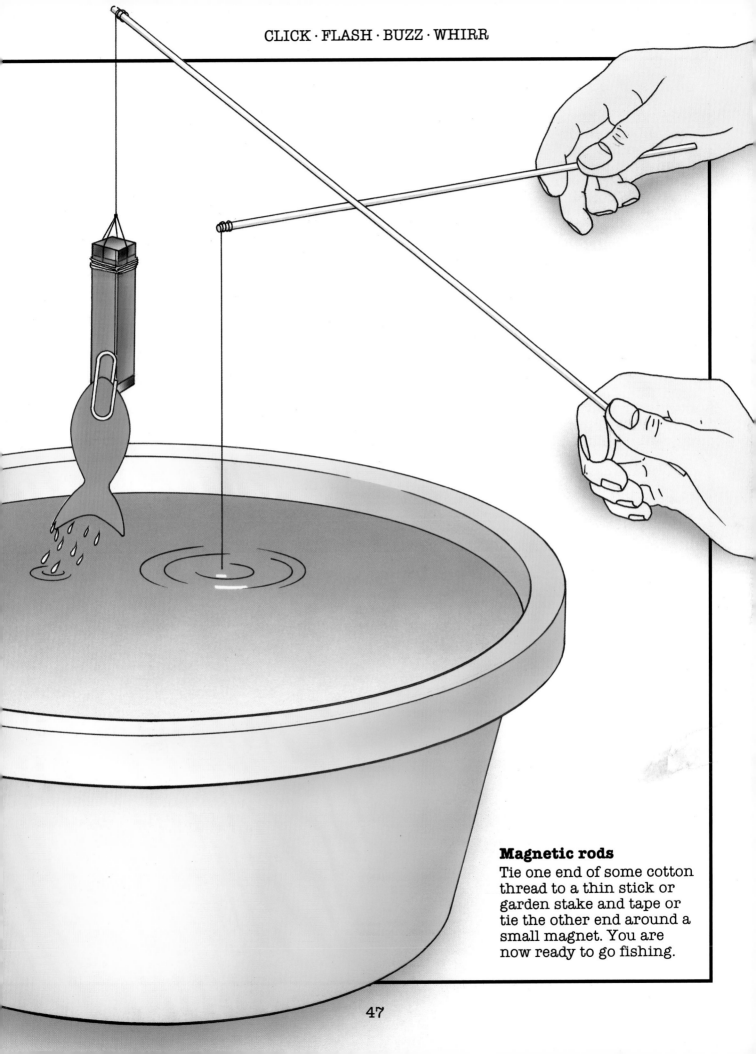

Magnetic rods

Tie one end of some cotton thread to a thin stick or garden stake and tape or tie the other end around a small magnet. You are now ready to go fishing.

47

MAGNETIC HOCKEY

To make this game you will need a large sheet of stiff cardboard which can be placed on four plastic cups turned upside down. This is the playing surface. Two steel washers make the players and a slice cut from a cork is the ball. Tape or tie a magnet to the end of two long sticks or garden stakes. These are used to move the players around the board. Finally cut out two goal shapes from cardboard and glue them at each end of the board.

How to play

Put the cork "puck" in the center and the two players in each goal line. Shout "go" and move your player towards the puck by sliding the magnetic stick under the board to attract the steel washer through the card. Goals are scored by pushing the puck with the washer over the goal line. Have a time limit of 10 minutes, changing ends at halftime.

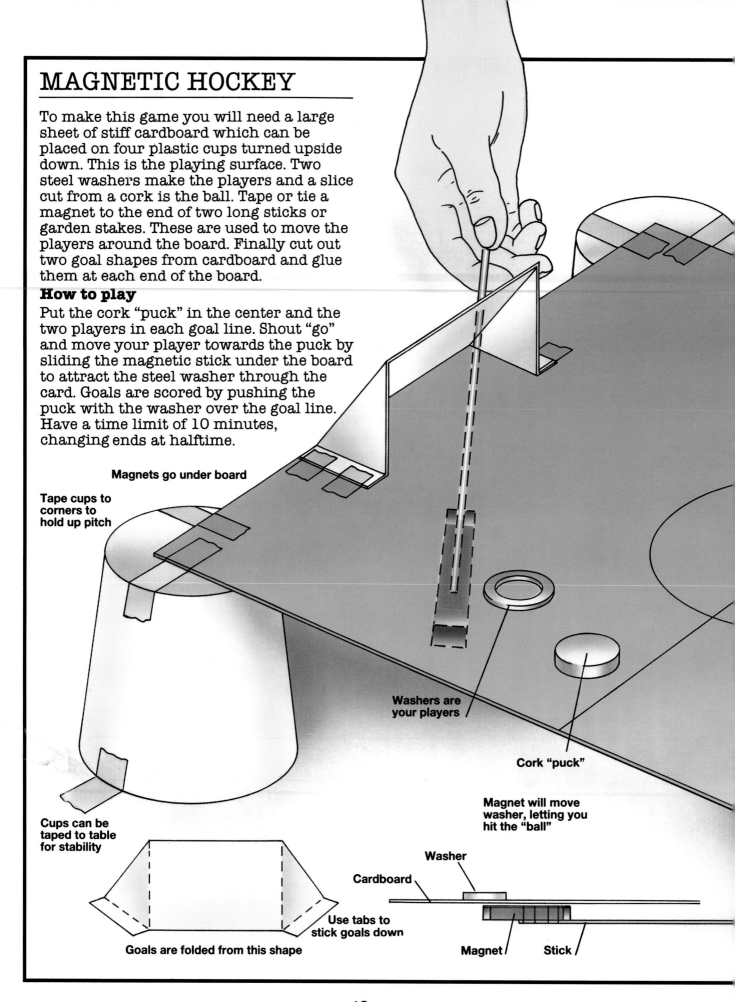

Magnets go under board

Tape cups to corners to hold up pitch

Cups can be taped to table for stability

Goals are folded from this shape

Use tabs to stick goals down

Washers are your players

Cork "puck"

Magnet will move washer, letting you hit the "ball"

Washer

Cardboard

Magnet

Stick

If you have enough magnets, more than two people can play

Color the washers to show the different teams

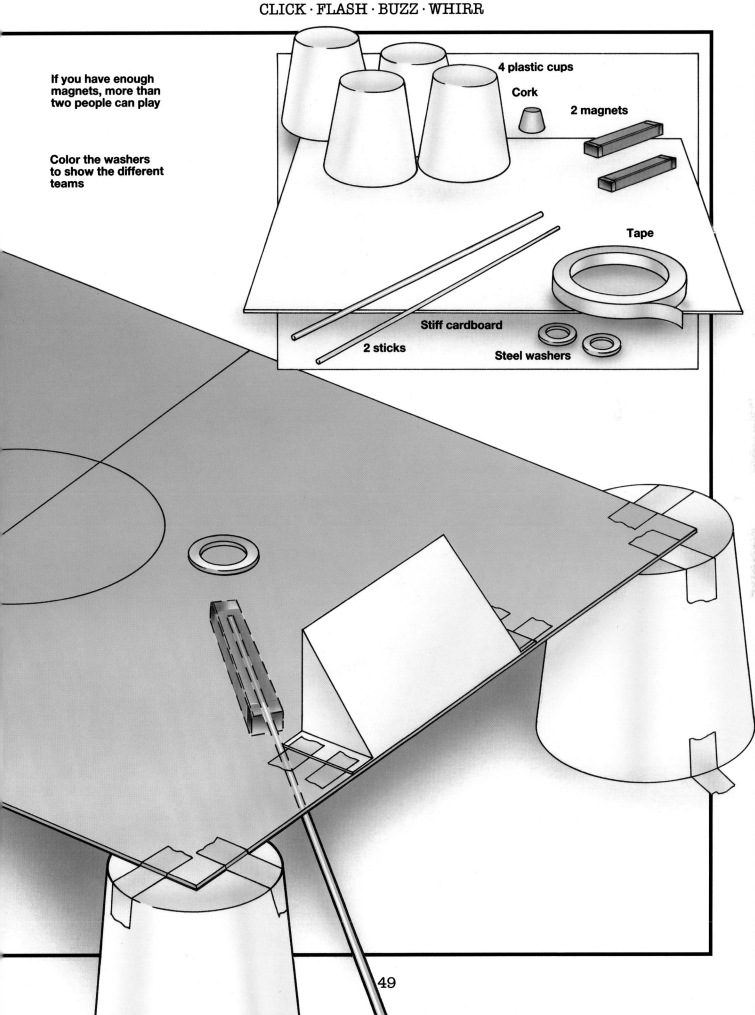

4 plastic cups

Cork

2 magnets

Tape

Stiff cardboard

2 sticks

Steel washers

49

STATIC ELECTRICITY

People have known about electricity for thousands of years. A philosopher and scientist called Thales who lived in Greece more than 2500 years ago found that if he rubbed pieces of amber with a cloth, the amber attracted small objects like pieces of dried grass and pepper grains. Thales had discovered static electricity, so called because it can naturally build up in things without flowing away as current electricity does through wires.

Thunder and lightning

The flashes of light and bangs that we see and hear during a thunder storm are caused by static electricity. They build up in clouds, then jump from one cloud to another or to the ground. The great heat of the electrical spark makes the air around it explode with a loud boom.

Thales the Greek

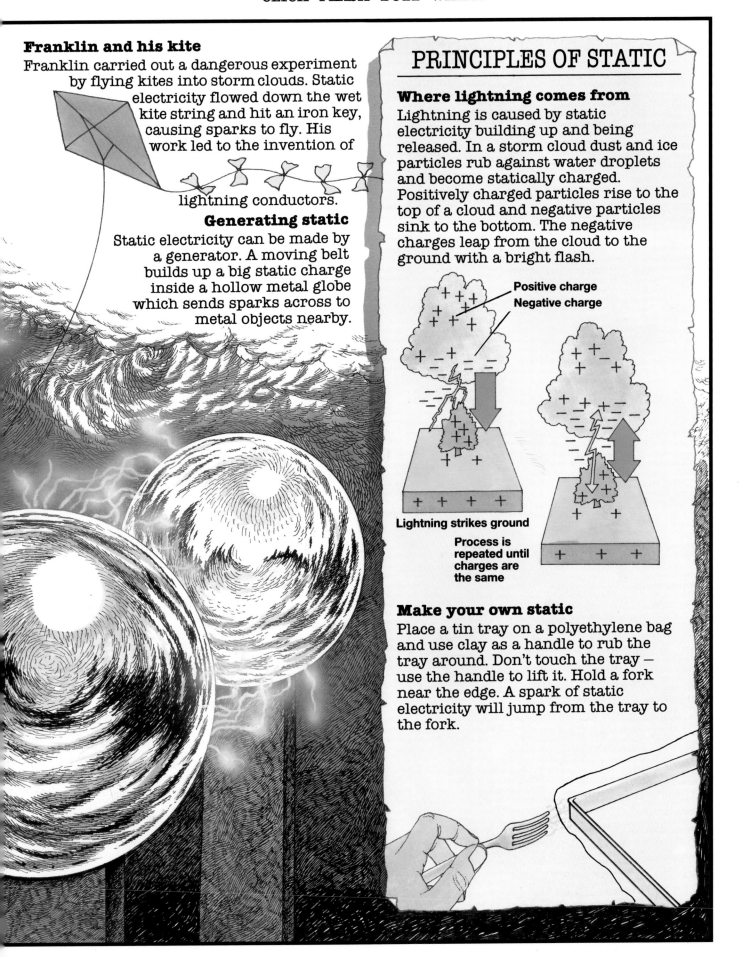

Franklin and his kite

Franklin carried out a dangerous experiment by flying kites into storm clouds. Static electricity flowed down the wet kite string and hit an iron key, causing sparks to fly. His work led to the invention of lightning conductors.

Generating static

Static electricity can be made by a generator. A moving belt builds up a big static charge inside a hollow metal globe which sends sparks across to metal objects nearby.

PRINCIPLES OF STATIC

Where lightning comes from

Lightning is caused by static electricity building up and being released. In a storm cloud dust and ice particles rub against water droplets and become statically charged. Positively charged particles rise to the top of a cloud and negative particles sink to the bottom. The negative charges leap from the cloud to the ground with a bright flash.

Positive charge
Negative charge

Lightning strikes ground

Process is repeated until charges are the same

Make your own static

Place a tin tray on a polyethylene bag and use clay as a handle to rub the tray around. Don't touch the tray — use the handle to lift it. Hold a fork near the edge. A spark of static electricity will jump from the tray to the fork.

MAKING STATIC ELECTRICITY

You do not need a laboratory or special equipment to make static electricity. Your experiment can be as simple as combing your hair. Run a plastic comb through your hair a number of times, or better still, rub it hard with a piece of woollen fabric. Then hold the comb near some tiny scraps of tissue paper. Rubbing the plastic builds up a static charge in the comb that attracts the tissue scraps. Try the same experiment with different plastic items. It works with pens, drinking straws and even balloons.

Rub objects on piece of wool

Which ones pick up scraps of tissue?

Use an assortment of plastics

STICKY BALLOONS

This is a very easy investigation that shows static electricity in action.

Blow up some balloons and rub them hard on a wool sweater. Hold the balloons against a wall. The balloons seem to stick on the wall as if by magic. What is happening is that the static charge on the balloons is different from the static charges on the wall, and opposite charges attract one another. The balloons are held against the wall until the charge gradually leaks away and the balloons slip to the floor.

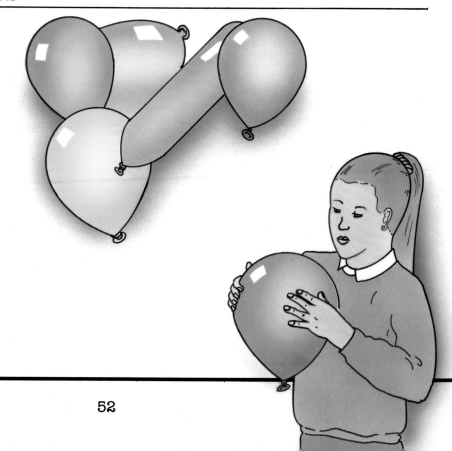

BENDING WATER

Use your balloon once more, but this time charge up your balloon by rubbing it hard on a wool sweater, then hold it near a running tap. The water is attracted towards the charged balloon and you can see the flow of water bending towards it.

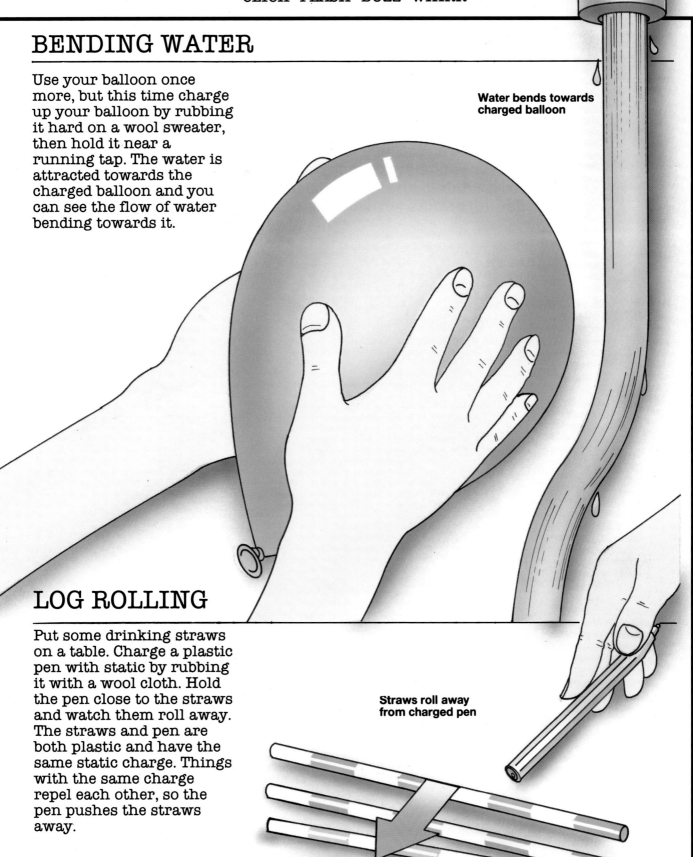

Water bends towards charged balloon

LOG ROLLING

Put some drinking straws on a table. Charge a plastic pen with static by rubbing it with a wool cloth. Hold the pen close to the straws and watch them roll away. The straws and pen are both plastic and have the same static charge. Things with the same charge repel each other, so the pen pushes the straws away.

Straws roll away from charged pen

MAKING ELECTRICITY

Life as we know it would be impossible without electricity. Think of all the things that it powers. Yet less than 200 years ago, electricity was a strange force that scientists didn't know how to use. It took the genius of many inventors to discover how electricity could be generated and many years of research before the first electric lights shone.

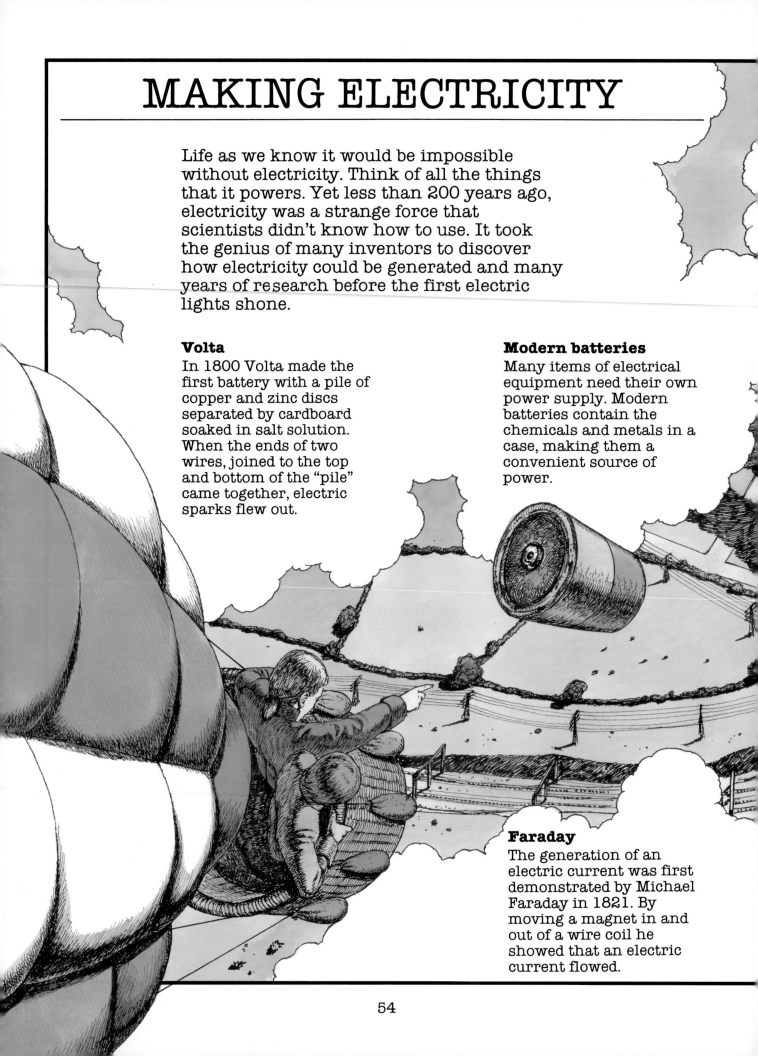

Volta

In 1800 Volta made the first battery with a pile of copper and zinc discs separated by cardboard soaked in salt solution. When the ends of two wires, joined to the top and bottom of the "pile" came together, electric sparks flew out.

Modern batteries

Many items of electrical equipment need their own power supply. Modern batteries contain the chemicals and metals in a case, making them a convenient source of power.

Faraday

The generation of an electric current was first demonstrated by Michael Faraday in 1821. By moving a magnet in and out of a wire coil he showed that an electric current flowed.

Power stations

Most of our electricity is generated in power stations. Fuel like oil or coal is burned to heat water and produce steam, which turns turbines. The turbines turn generators which make electricity.

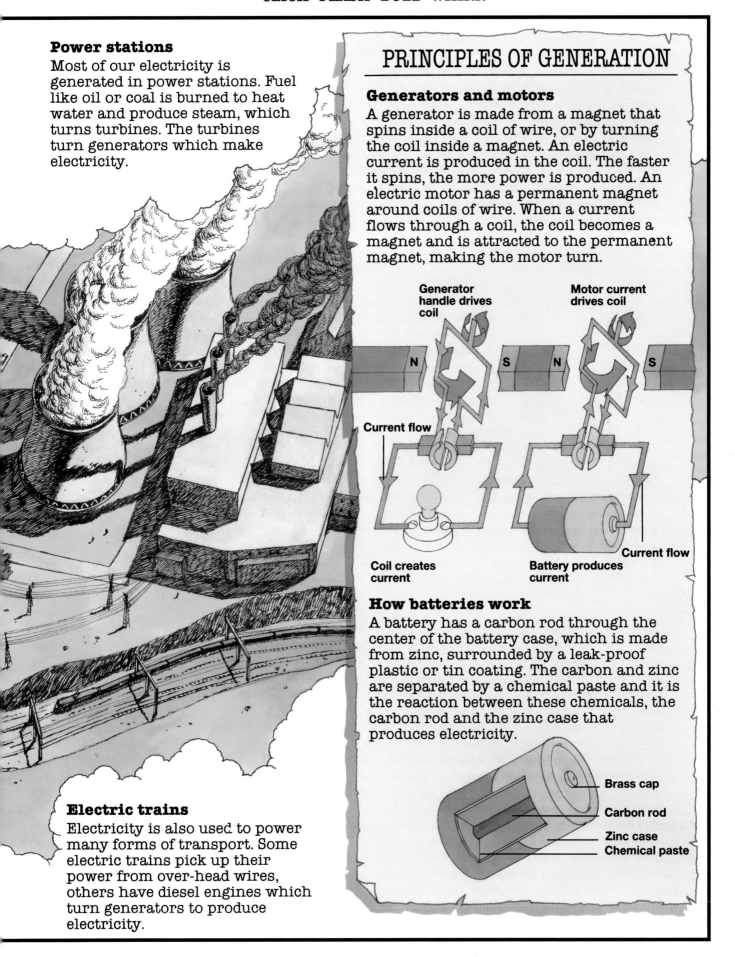

PRINCIPLES OF GENERATION

Generators and motors

A generator is made from a magnet that spins inside a coil of wire, or by turning the coil inside a magnet. An electric current is produced in the coil. The faster it spins, the more power is produced. An electric motor has a permanent magnet around coils of wire. When a current flows through a coil, the coil becomes a magnet and is attracted to the permanent magnet, making the motor turn.

Generator handle drives coil

Motor current drives coil

N   S   N   S

Current flow

Current flow

Coil creates current

Battery produces current

How batteries work

A battery has a carbon rod through the center of the battery case, which is made from zinc, surrounded by a leak-proof plastic or tin coating. The carbon and zinc are separated by a chemical paste and it is the reaction between these chemicals, the carbon rod and the zinc case that produces electricity.

Brass cap

Carbon rod

Zinc case
Chemical paste

Electric trains

Electricity is also used to power many forms of transport. Some electric trains pick up their power from over-head wires, others have diesel engines which turn generators to produce electricity.

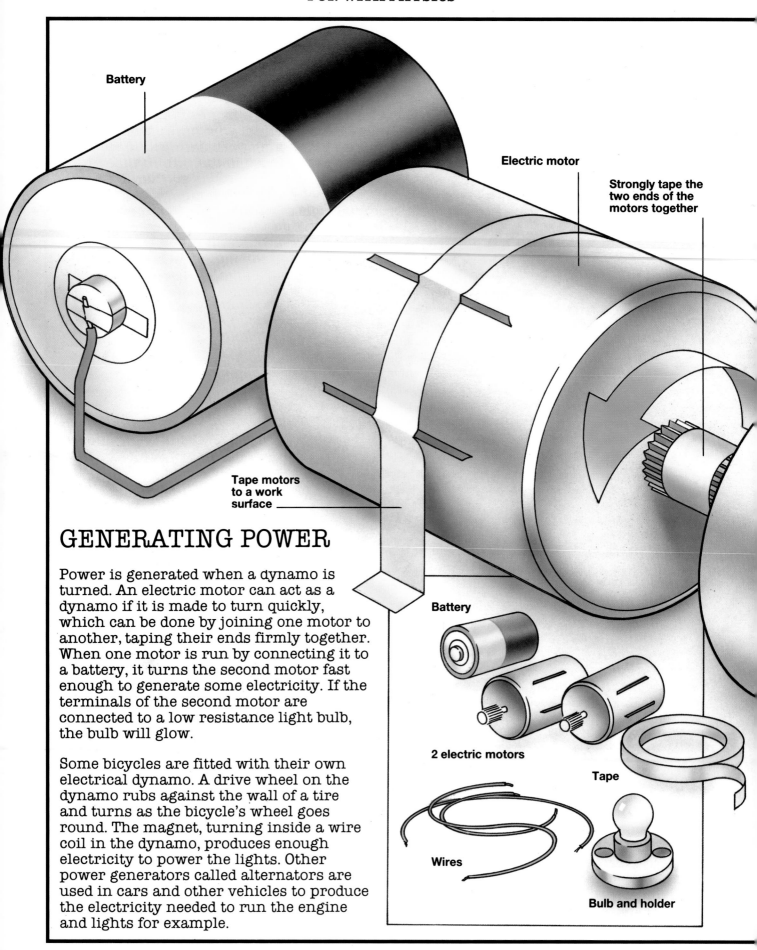

Battery

Electric motor

Strongly tape the
two ends of the
motors together

Tape motors
to a work
surface

Battery

2 electric motors

Tape

Wires

Bulb and holder

GENERATING POWER

Power is generated when a dynamo is
turned. An electric motor can act as a
dynamo if it is made to turn quickly,
which can be done by joining one motor to
another, taping their ends firmly together.
When one motor is run by connecting it to
a battery, it turns the second motor fast
enough to generate some electricity. If the
terminals of the second motor are
connected to a low resistance light bulb,
the bulb will glow.

Some bicycles are fitted with their own
electrical dynamo. A drive wheel on the
dynamo rubs against the wall of a tire
and turns as the bicycle's wheel goes
round. The magnet, turning inside a wire
coil in the dynamo, produces enough
electricity to power the lights. Other
power generators called alternators are
used in cars and other vehicles to produce
the electricity needed to run the engine
and lights for example.

TURNING POWER

Electricity can be made whenever a generator can be turned fast enough. There are many ways in which this can be done and engineers are experimenting with different methods so that we will not run out of precious fossil fuels like coal and oil. One method that is becoming more and more important uses the power of water trapped behind dams to turn turbine blades. Even the force of the wind can be used to turn massive blades connected to generators.

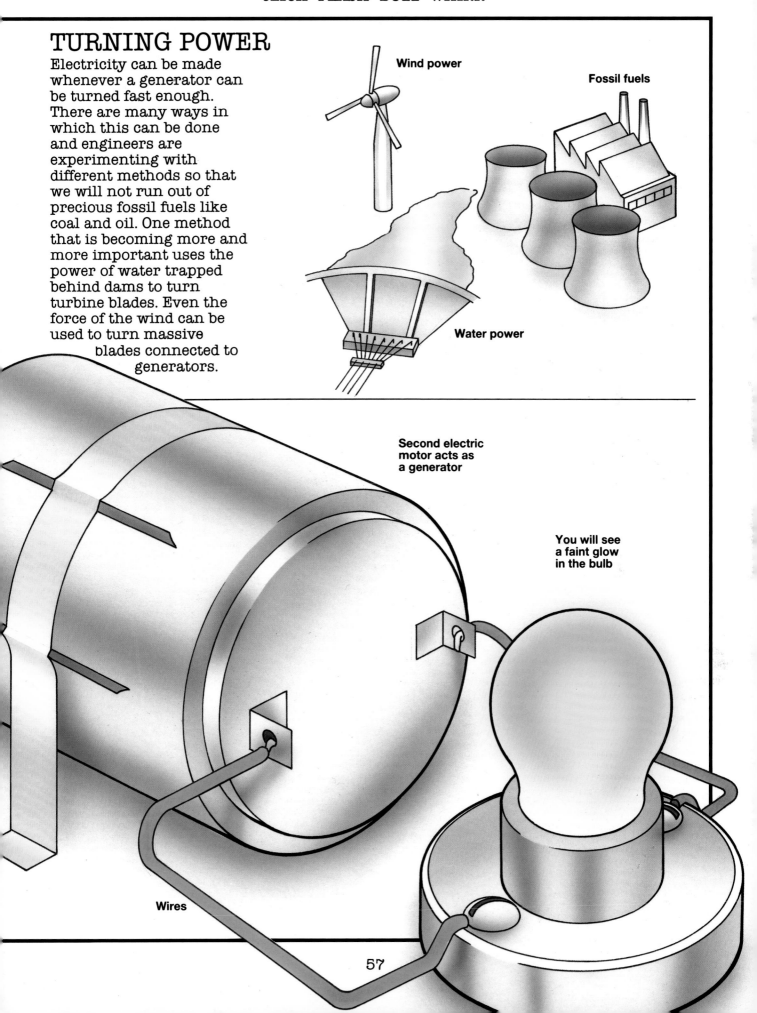

Wind power

Fossil fuels

Water power

Second electric motor acts as a generator

You will see a faint glow in the bulb

Wires

CIRCUITS

Electricity can only do useful work when it is made to flow in a complete circuit connected to a bulb or motor, or inside the circuitry of televisions, radios and so on. Most electrical equipment uses high power from household current. This is very dangerous and you should **never** play with electricity from sockets. Experiments into electrical circuits can be safely carried out using batteries.

**NEVER USE
ELECTRICITY
FROM SOCKETS
ONLY BATTERIES**

Switches

A switch is a gap in a circuit that electricity cannot cross. Switches allow us to have control over the flow of electricity because they can change the direction of the flow or start and stop it.

Conductors and insulators

Electricity needs a pathway to flow through. Materials that allow it to flow, such as copper, are conductors. Those that "stop" electrical flow, like plastic, rubber and pottery, are insulators.

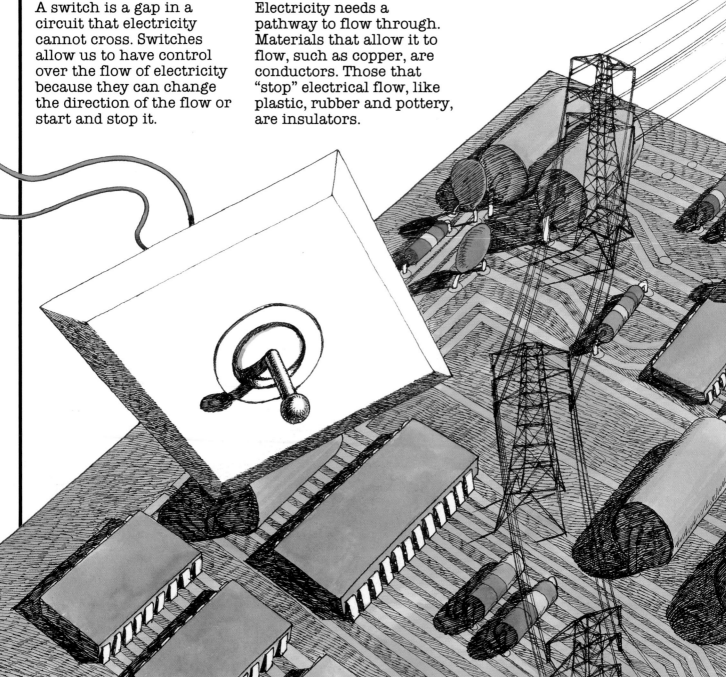

PRINCIPLES OF ELECTRON FLOW

Everything is made from tiny particles, called atoms. The center of an atom is called the nucleus, which is surrounded by electrons. In good electrical conductors, like metals, the outer electrons are free to move. When a conductor is connected to a power supply, these electrons are driven in one direction. The movement of electrons around a circuit is called an electric current. Power supplied by a battery pushes the electrons in one direction only. This is direct current. Electricity generation in a power station flows first one way, then the other, changing about 50 times a second. This is alternating current.

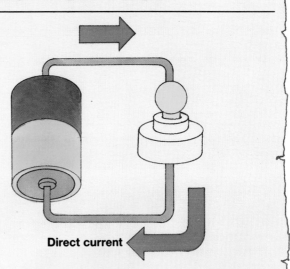

Direct current

Valves

Valves allow the movement of electrons in one direction. Special valves can increase the flow and were used in early televisions and radios.

Circuit boards

Electrical equipment became simpler and smaller with the invention of transistors and integrated circuits. Many parts could be put together on circuit boards where metal strips were used to make the necessary connections.

Micro-chips

One tiny micro-chip has the same power as thousands of transistors and other components, on integrated circuits the size of a drawing pin head.

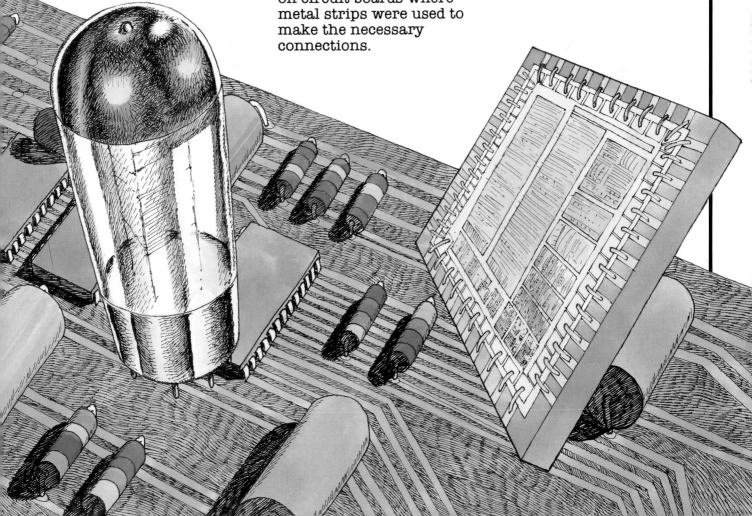

WHAT CONDUCTS?

Take a battery, a bulb holder, two wires and lots of different objects. Include coins, cork, glass, china, keys, stone, plastics, wood, rubber, cutlery and card. Make a circuit as shown and touch the two bare ends of the wires together. This completes the circuit and the bulb will shine. Take each object and touch the wires to it. If the bulb lights, electricity is able to flow through the object, so it is a conductor. If the bulb does not light, the object stops the current, so it must be an insulator. Separate the conductors and insulators. What are the conductors made from?

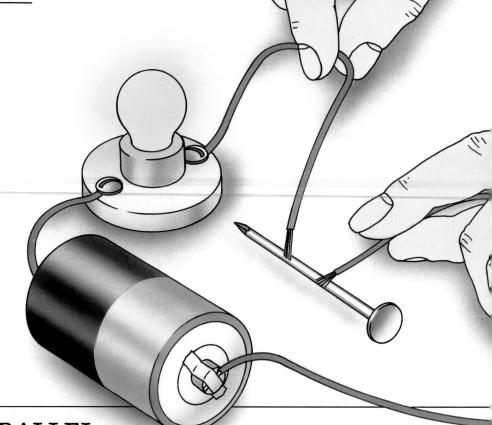

SERIES AND PARALLEL

It is quite easy to get a bulb to light using one battery, but can you still get it to light with two batteries? Try different arrangements of batteries. Which ones work and which do not? Can you make two or even three bulbs light from only one battery? Try different arrangements of wires to see if you can get them to shine brightly. Two different ways are shown for you. Try making the "series" circuit. Are the bulbs bright? What happens when you unscrew one of the bulbs? Now try making the "parallel" circuit. Is there a difference in the brightness of the bulbs? What happens now when one is unscrewed?

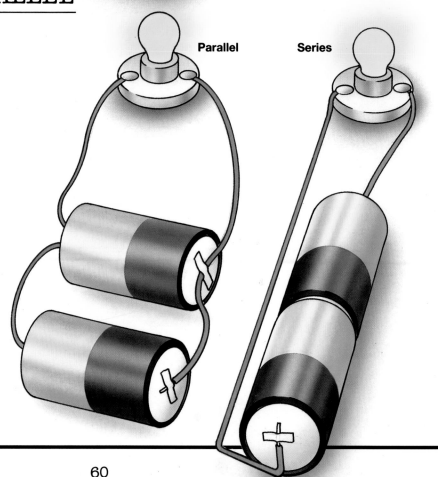

Parallel   **Series**

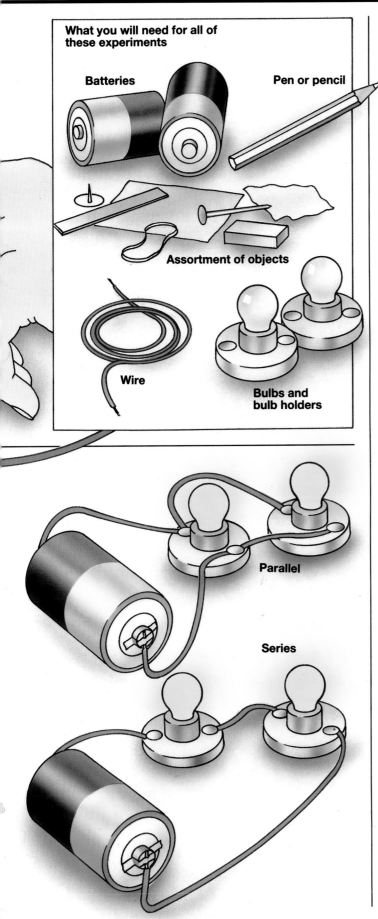

What you will need for all of these experiments

Batteries

Pen or pencil

Assortment of objects

Wire

Bulbs and bulb holders

Parallel

Series

RESISTANCE COIL

Get a long length of thin bare wire. Wrap it tightly around a pencil as many times as you can. Carefully connect your coil into the circuit as shown. The wire from the bulb holder can be touched at any point on the coil. Watch the brightness of the bulb each time. Thin wire has a high resistance to an electric flow. If you touch the coil near its start, where only a few turns have been made, the resistance is not as great and the bulb is bright. Further along the coil are more turns of wire, so there is higher resistance which means the bulb gets dimmer.

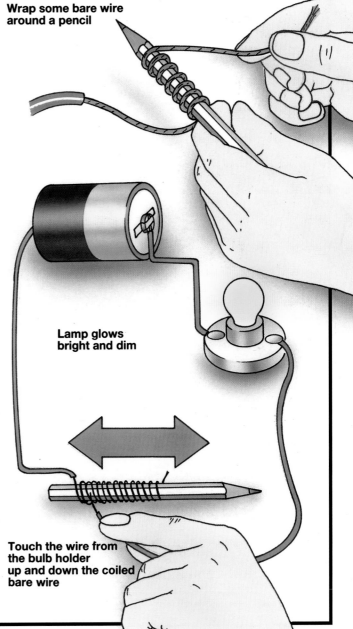

Wrap some bare wire around a pencil

Lamp glows bright and dim

Touch the wire from the bulb holder up and down the coiled bare wire

SWITCHES

A switch is nothing more than a gap in a circuit. Simple switches can be made easily using conductors such as paperclips and tin foil. These will need to be mounted on a suitable base such as thick cardboard or wood. Make a simple circuit with a bulb and battery. Carefully cut one of your wires, trapping its bare ends under thumbtacks pushed into a soft wood base. Also trap a paperclip or a strip of tin foil under one of the thumbtacks. When the paperclip or foil touches the other thumbtack, the circuit is completed and the bulb will shine.

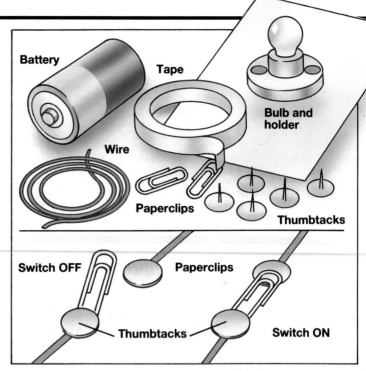

Battery Tape Bulb and holder Wire Paperclips Thumbtacks

Switch OFF Paperclips Thumbtacks Switch ON

Double switch

This works in the same way as a landing light when one switch is downstairs and the other is upstairs. The light can be turned on or off from either switch.

Simple switch

The single paperclip switch can be either a slide switch, moving the clip from side to side to turn on and off, or a push-to-make switch. Bend the end of the paperclip up at an angle above the second drawing pin. When the clip is pushed down the light comes on, and when it is released it springs back up.

CODED MESSAGES

In 1843, Samuel Morse set up an
experimental telegraph in the United
States, sending coded messages of
electrical pulses along wires between
telegraph offices. The Morse code was
taken up by all countries for sending
messages using a pattern of short and
long electrical signals, sounds or light
flashes.

How to make it

Make your own Morse code sender
and receiver by building two bulb
circuits as shown, with push-to-make
switches. Connect the two together
with a long length of double wire.

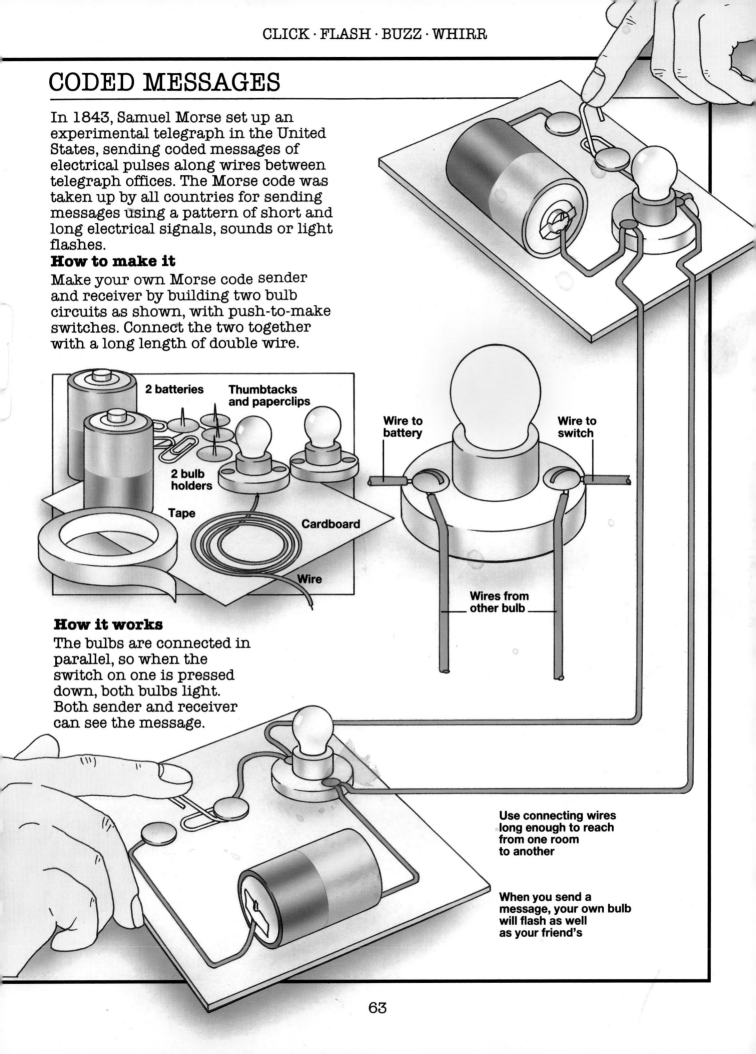

2 batteries

**Thumbtacks
and paperclips**

**2 bulb
holders**

Tape

Cardboard

Wire

**Wire to
battery**

**Wire to
switch**

**Wires from
other bulb**

How it works

The bulbs are connected in
parallel, so when the
switch on one is pressed
down, both bulbs light.
Both sender and receiver
can see the message.

**Use connecting wires
long enough to reach
from one room
to another**

**When you send a
message, your own bulb
will flash as well
as your friend's**

ELECTROMAGNETISM

Magnetism and electricity are closely
linked. In 1819 the Danish physicist, Hans
Christian Oersted, discovered the magnetic
effect of an electric current. He noticed that
a compass needle near a wire carrying an
electric current swung round, lining up at
right angles to the wire. Electricity, flowing
in a coil, can magnetize a steel bar and both
generators and motors use electromagnetic
effects to produce power or to turn.

On cranes

Large electromagnets are
often used in scrap-metal
yards to lift iron and steel.
They can also be used to
sort metals, removing only
the ferrous metals (iron
and steel), and leaving
behind non-magnetic
metals and scrap.

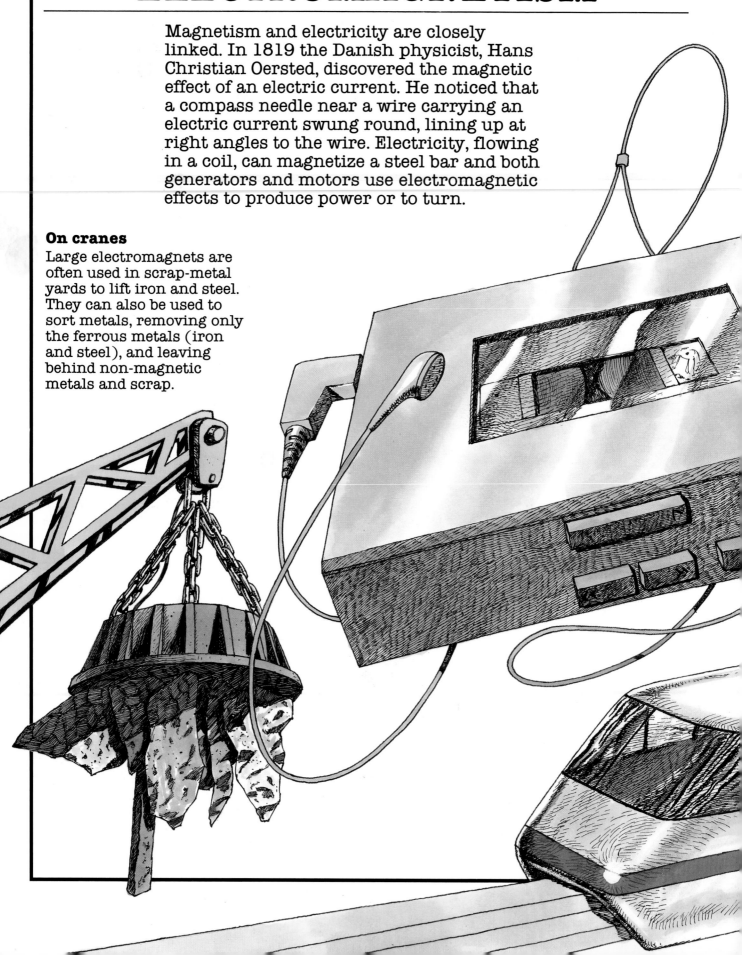

PRINCIPLES OF ELECTROMAGNETISM

A current in a coil of wire sets up a strong magnetic force. If a steel or iron bar is placed in the coil, the magnetic domains in the metal line up with the magnetic lines of force running through the coil. With a steel bar the effect on the bar is permanent and the bar remains magnetized even when the current is turned off. If the bar has a soft iron core the effects only last as long as the current is switched on. So an electromagnet with a soft iron core picks up iron and steel only when the current is on.

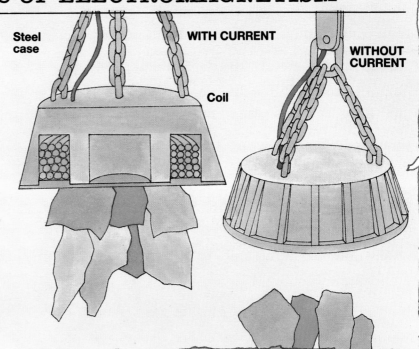

Steel case

Coil

WITH CURRENT

WITHOUT CURRENT

Sound recording

Tape in video and cassette recorders is covered with a special metal. The recording head changes the pattern of the tape's metallic particles, which the play-back head then turns into signals that are played through speakers.

Floating trains

Experimental trains and tracks have electromagnets whose north poles are touching. The magnets repel one another pushing the train clear of the track. As there is almost no friction the train can move easily and go very fast.

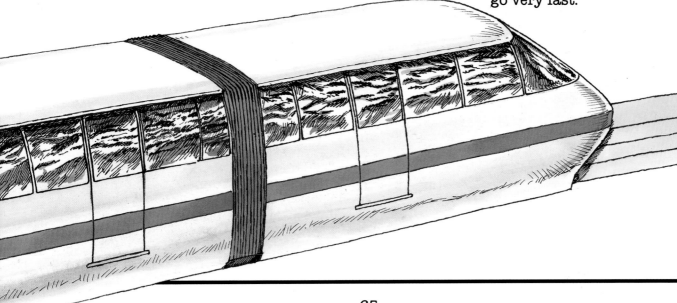

ELECTROMAGNET

To make your own electromagnet you need a long iron nail, some thin insulated wire, a battery and a switch. Wind the wire tightly and evenly around the nail. Use tape to hold the coil in place. Put as many turns on the nail as you can — the more turns, the stronger the magnet. Make a simple switch from thumbtacks and a paperclip and connect your coil to the switch and the battery. Complete the circuit with a wire from the other side of the switch to the battery and switch on. With the current flowing, dip the end of the nail into a pile of small nails, pins or paperclips. Lift your electromagnet clear and see how many you have picked up.

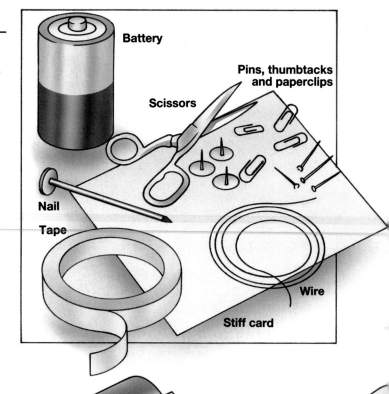

Battery

Pins, thumbtacks and paperclips

Scissors

Nail

Tape

Wire

Stiff card

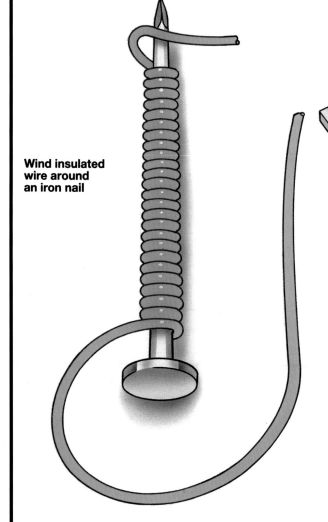

Wind insulated wire around an iron nail

How it works

When the current is on, the coil sets up a magnetic field that magnetizes the nail. The magnetic nail attracts the pins and clips. When the power is turned off, the nail loses its magnetism and the pins and clips fall off.

MAGNETIC CRANE

You can use your electromagnet in a super model crane. Cut and fold the crane's jib or arm from a piece of stiff cardboard and attach it to the box by pushing the point of the pencil through the side of the box, through the jib, then out the other side. The point of a pair of scissors can be used to make the holes. Take care with these. Tape the battery inside the box and make a switch by pushing paper fasteners through its side. Trap a paperclip under one of them and secure the wires around the "wings" of the fasteners once they have been opened out.

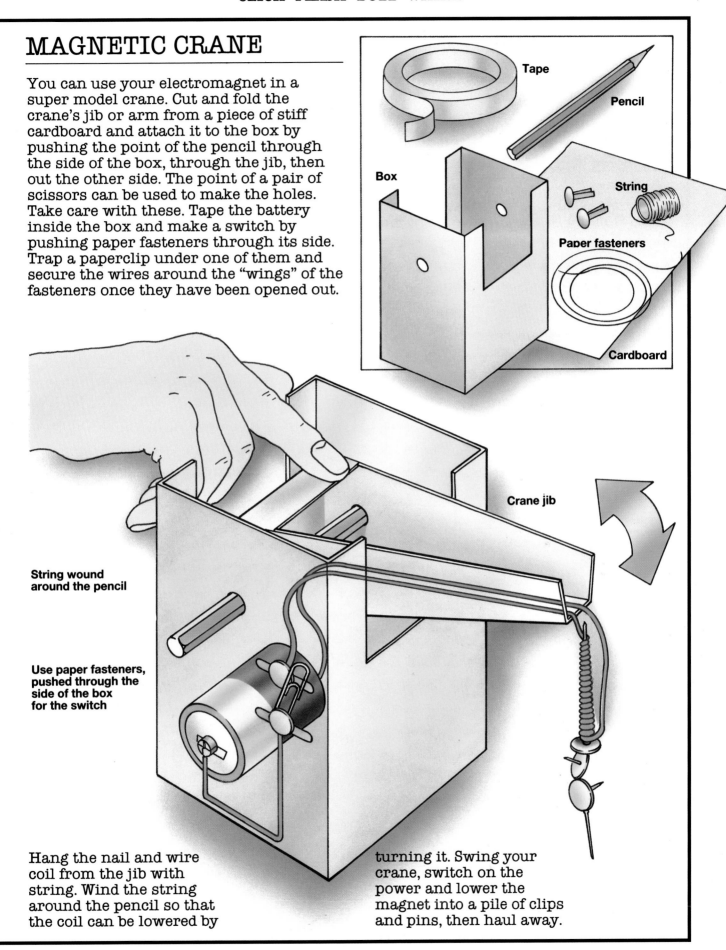

Tape

Pencil

Box

String

Paper fasteners

Cardboard

Crane jib

String wound around the pencil

Use paper fasteners, pushed through the side of the box for the switch

Hang the nail and wire coil from the jib with string. Wind the string around the pencil so that the coil can be lowered by turning it. Swing your crane, switch on the power and lower the magnet into a pile of clips and pins, then haul away.

MAGNETIC PATTERNS

Tape your electromagnet under a sheet of stiff cardboard supported on four cups turned upside down. Sprinkle some iron filings over the sheet and switch on. Tap the card gently and you will see the iron filings taking up positions along magnetic lines of force. You can make a permanent record of these patterns by carefully spraying the surface of the card with an ozone-friendly spray adhesive to fix the filings in place.

Amaze your friends

You can place small metal objects such as clips and pins on the surface of the board. As you turn your electromagnet on and off, hidden from sight below the board, the metal objects will dance as they are moved by the magnet.

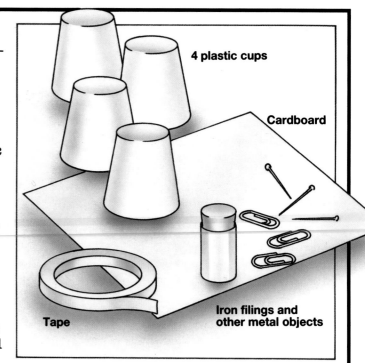

4 plastic cups

Cardboard

Iron filings and other metal objects

Tape

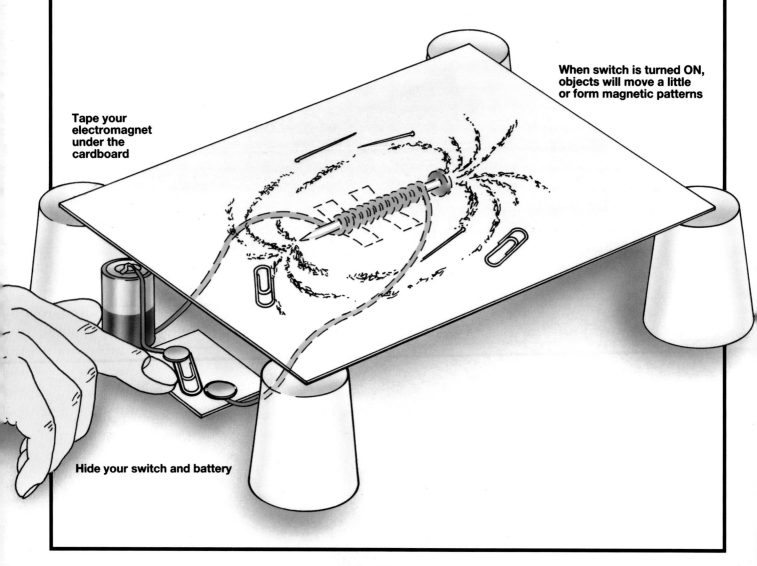

When switch is turned ON, objects will move a little or form magnetic patterns

Tape your electromagnet under the cardboard

Hide your switch and battery

3

SIZZLE
FREEZE
BUBBLE
GLOW

THE SCIENCE AND TECHNOLOGY OF HEAT, LIGHT AND CHEMICAL ENERGY

LIGHT AND SHADOWS

Our Earth seems a very large place. But it is only a small planet and it is a tiny scrap compared to the Sun. The Sun looks small because it is about 93 million miles away, but without it there could be no Earth. The Sun gives us warmth, light and life. All the energy that is stored in food and fuel came originally from the Sun.

The burning Sun
The Sun is the star at the center of our solar system. It is a huge ball of glowing gas with a diameter of 864,000 miles (Earth's diameter is 7,926 miles), yet compared with some stars our Sun is only a dwarf star.

Mirrors and reflection
Light travels in straight lines. It can be bounced back or reflected by mirrors of silvered glass, polished metal or smooth surfaces that reflect light.

Shadows
When a solid object lies in the path of a beam of light, blocking the light, the object is called "opaque." Opaque objects stop the light, leaving a darkened patch called a shadow.

Lenses and refraction
Light can also be bent or refracted using lenses. Lenses that focus light to a point are called "converging" lenses, those that spread light are "diverging" lenses.

Light and color
We see colors when light shines on things. It is hard to make out colors in the dark, as very little light is reflected from them into our eyes.

PRINCIPLES OF LIGHT

What is light?
We see when light reflects off things into our eyes. Light is a form of energy called electromagnetic radiation, made up of electrically charged particles. The total spectrum of radiation includes X-rays, ultraviolet rays, infrared and radio waves. They each have a different wavelength, or speed of vibration (frequency). Our eyes only react to a very small group of wavelengths that makes visible light.

Radio waves	Micro-waves	Infrared rays	Visible light	Ultraviolet rays	X-rays	Gamma rays

| | | Black | White | Black | What we see |

The Sun's energy
Light from the Sun is a mixture of seven colors that we see as white light. Huge amounts of heat and light energy are produced from nuclear reactions deep inside the Sun. Temperatures on the surface can reach 10,000 degrees Fahrenheit (°F).

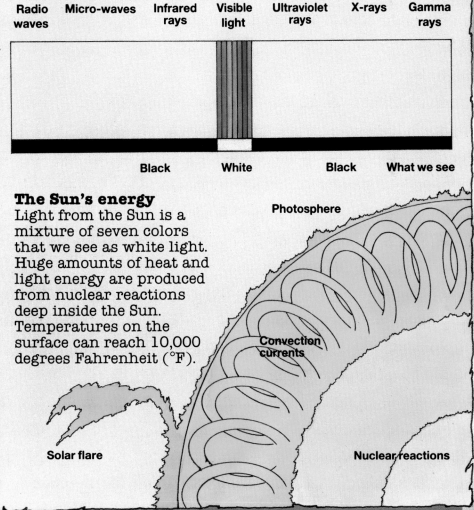

Photosphere

Convection currents

Solar flare

Nuclear reactions

MAKE YOUR OWN SUNDIAL

Set up your sundial on a window sill in a south-facing room early in the morning. You will need a watch and a pen. When your watch shows a whole hour (like 8 a.m.), draw a line passing through the center of the pencil's shadow and write the time beside it. Do this every hour, on the hour, until late into the afternoon. Notice how the length of the shadow changes during the day.

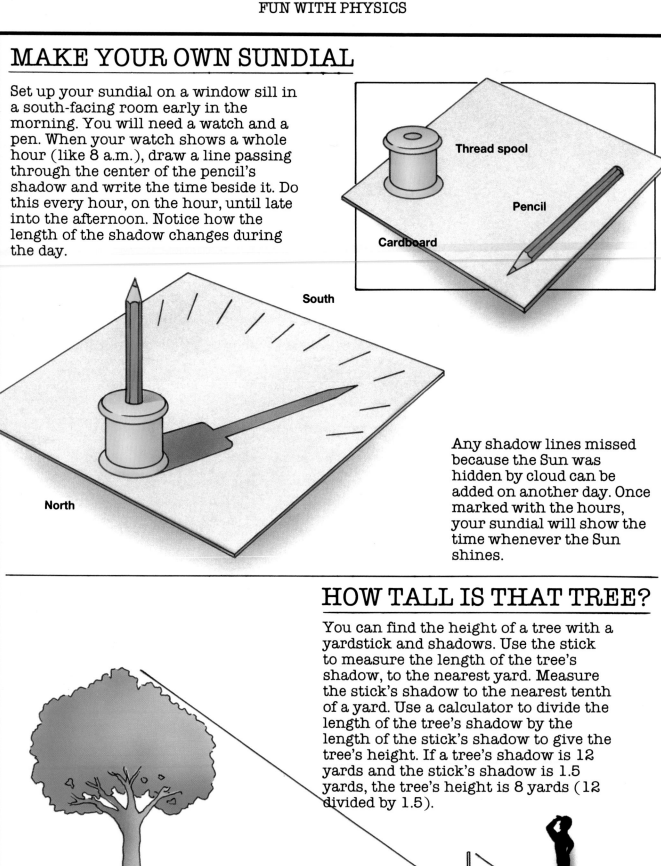

Thread spool

Pencil

Cardboard

South

North

Any shadow lines missed because the Sun was hidden by cloud can be added on another day. Once marked with the hours, your sundial will show the time whenever the Sun shines.

HOW TALL IS THAT TREE?

You can find the height of a tree with a yardstick and shadows. Use the stick to measure the length of the tree's shadow, to the nearest yard. Measure the stick's shadow to the nearest tenth of a yard. Use a calculator to divide the length of the tree's shadow by the length of the stick's shadow to give the tree's height. If a tree's shadow is 12 yards and the stick's shadow is 1.5 yards, the tree's height is 8 yards (12 divided by 1.5).

Length of tree's shadow

Length of stick's shadow

PLAYING WITH SHADOWS

Look for your shadow on a bright, sunny day. Can you run away and leave it behind? Shadow box with a friend's shadow, without touching each other. Use a torch or table lamp in a dark room to make shadows on a wall. Watch what happens to their size if you move things towards or away from the light.

Hand shadows

Try making animal shapes with your hands, like a bird that flaps its wings. Use your finger tips to make the head of a dog.

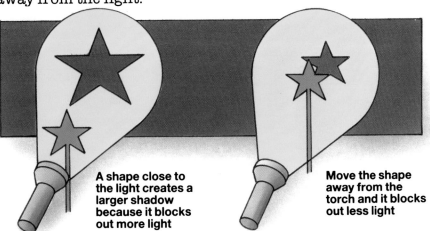

A shape close to the light creates a larger shadow because it blocks out more light

Move the shape away from the torch and it blocks out less light

SHADOW PUPPET THEATER

Light source

Use a lamp to light up a screen made from an old cotton sheet or tablecloth. Carefully cut out scenery and character shapes from cardboard. Pin the scenery to the sheet. Tape your characters to thin sticks, like garden stakes, so they can move across the screen. Make up a play to entertain your friends and family. You can make your shadow puppets even more realistic by having moving parts held together with paper fasteners.

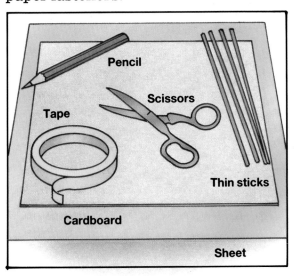

Pencil

Scissors

Tape

Thin sticks

Cardboard

Sheet

The closer the puppet is to the sheet, the sharper the shadow will be

You can move your characters more if you use paper fasteners to hold them together

Stick

Cardboard

LIGHT TRAVELS IN STRAIGHT LINES

You can show that light travels in a straight line
if you put a tiny light, like a small bulb or a candle,
at one end of a table. (Have an adult stand a candle
in a tray full of sand or set securely on a saucer.)
Make a small hole in a square of cardboard and
set the square in clay so that you can see the light
through the hole. Line up the hole in a second
square so that you can still see the bulb or flame.
Set up a third square, too, so that you can see the
light through all three holes. Always blow out the
candle after your investigation.

**DON'T USE A
CANDLE WITHOUT
HELP FROM AN
ADULT**

**Line all three cards
up until you can see
the flame**

Card

Modelling clay

**The path of the light
is a straight line**

Push a piece of thin dowel
rod through the center of
the holes. It will point
directly at the light,
proving that the light's
path is perfectly straight.

MAKE YOUR OWN LIGHT-RAY BOX

Many interesting experiments can be done
using a light box. The light from a flashlight
shines through the slit in the box. Put your
light box on a table in a darkened room and
set a mirror in front of it. Watch how the
angle between the light beam and its
reflection changes as you move the mirror.

Mirror

Flashlight

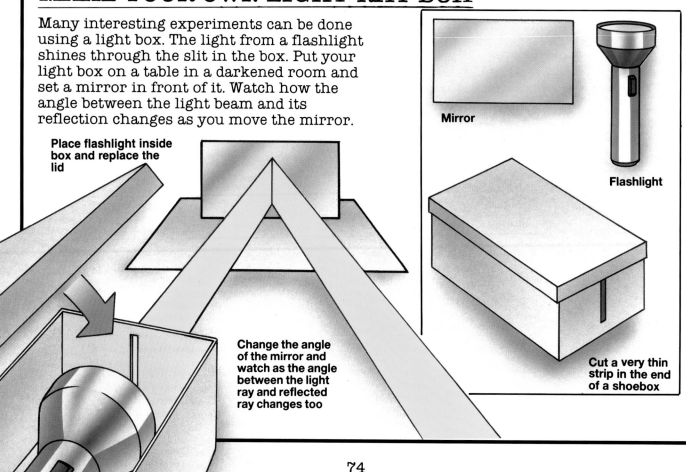

**Place flashlight inside
box and replace the
lid**

**Change the angle
of the mirror and
watch as the angle
between the light
ray and reflected
ray changes too**

**Cut a very thin
strip in the end
of a shoebox**

MAKE YOUR OWN PERISCOPE

Can you see around a corner or over a wall without being seen? Submarines use periscopes to see what is above them without having to come to the surface. You can make a periscope from cardboard with slits cut into the sides to take two small mirrors. You must set the angle of the mirror slits at 45° (half a right angle), so that the light will be reflected through the periscope tube.

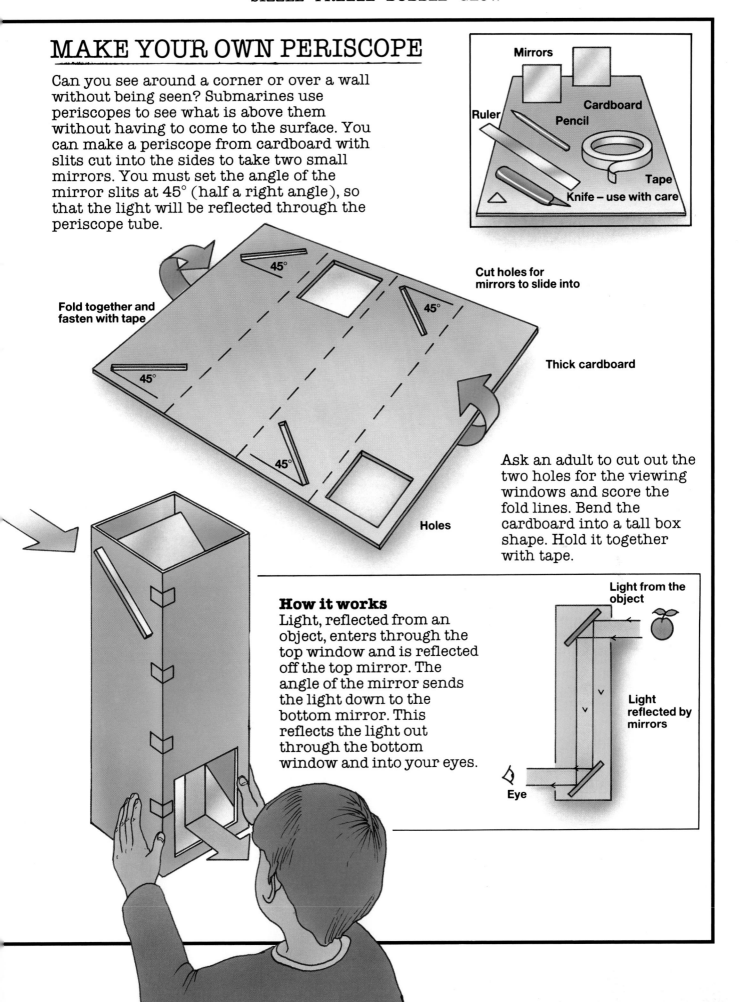

Mirrors
Cardboard
Ruler
Pencil
Tape
Knife – use with care

45°
45°
45°
45°

Fold together and fasten with tape

Cut holes for mirrors to slide into

Thick cardboard

Holes

Ask an adult to cut out the two holes for the viewing windows and score the fold lines. Bend the cardboard into a tall box shape. Hold it together with tape.

How it works
Light, reflected from an object, enters through the top window and is reflected off the top mirror. The angle of the mirror sends the light down to the bottom mirror. This reflects the light out through the bottom window and into your eyes.

Light from the object
Light reflected by mirrors
Eye

MAKE YOUR OWN KALEIDOSCOPE

Kaleidoscopes are very old toys that use mirrors to produce colorful, changing patterns. You can make your own by taping together three small mirrors into a triangle. Hold the mirrors over a collection of small beads or brightly colored tissue paper scraps and look down inside the mirrors. To change the pattern, find a transparent dish or jar that the mirrors can stand in. Hold the dish and the mirrors up, so that light can shine in through the bottom, and slowly turn the dish as you look down. The effect is even better if you can hold your kaleidoscope at an angle so that the beads move inside the mirrors.

Multiple reflections

Tape two small mirrors together and stand them on a table top. Place a small toy between them and count the number of reflections that you can see. Make the angle larger, by moving the mirrors apart. Is the number of reflections more or less? Now move the mirrors closer together. Watch what happens to the number of reflections.

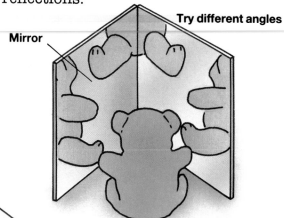

Try different angles

Mirror

Tape the backs of the three mirrors together

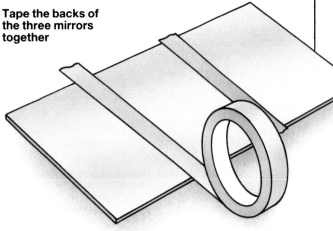

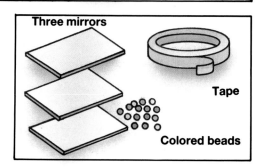

Three mirrors

Tape

Colored beads

Then fold inwards

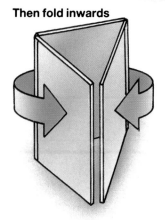

Multiple images appear

The beads or tissue scraps are reflected backwards and forwards between the mirrors, building complex patterns of colored light. As you turn the dish the beads roll around, moving about to form new patterns.

SPOON MIRRORS

Have you ever been to a fairground hall of mirrors? The strangely shaped mirrors make you look very odd indeed. Any bendy mirror will distort or alter your shape. You may be lucky enough to have a plastic mirror that you can bend gently to change to the shape of your reflection. If not, you can use a large spoon. Look into the bowl of a very shiny spoon. This is like a concave or hollowed out mirror. What is strange about your image? Now turn the spoon around and look at your reflection in the back of the spoon. This is a convex mirror. How do you look now?

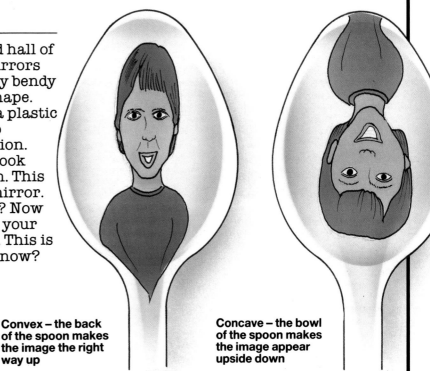

Convex – the back of the spoon makes the image the right way up

Concave – the bowl of the spoon makes the image appear upside down

MYSTERY MONEY

If you look at fish in a pool they seem quite close. When you dip your hand in to touch them you find they are deeper than you thought. Your eyes are being fooled by an illusion created by the water. Light rays bend as they pass from water into air and make you think that the water is shallower than it really is.

Bendy pencil
Light rays change direction as they pass from water to air, so any object in water seems to bend. A pencil seems to be bent at the place where it enters the water.

Put a coin in a bowl. Look over the rim, slowly bend down until the coin disappears from view. Get someone to pour water slowly into the bowl. As it fills the coin will reappear.

Line of sight

The light reflected from the coin is bent towards you as it passes from the water into the air. So you can now see the coin, even though the rim hid it before.

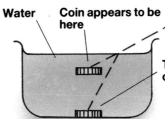

Water   **Coin appears to be here**

True position of the coin

NEAR AND FAR

Lenses bend or refract light, changing the image that we see. Someone who finds it hard to see objects close up (this is called being long-sighted), wears glasses fitted with convex or magnifying lenses to re-focus the light onto the retina at the back of the eyes. People who cannot focus on far-off objects (they are short-sighted), wear glasses with concave lenses. You can see the effects of these different lenses by holding them in the path of a series of light rays shining through slots cut into the end of a shoebox. Set up the light box in a darkened room and observe how the light rays are bent, or refracted, through the lenses.

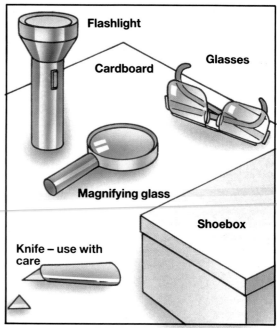

Flashlight

Cardboard

Glasses

Magnifying glass

Shoebox

Knife – use with care

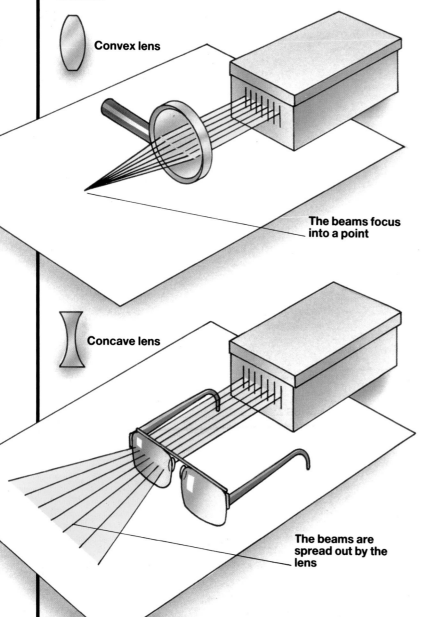

Convex lens

The beams focus into a point

Concave lens

The beams are spread out by the lens

Water drop lens

Water has a "skin" caused by tension in the surface that pulls small drops of water into a rounded shape. Because of this, water drops can act like tiny convex lenses. Put a drop of water on a thin piece of transparent plastic or glass and place it on a picture. You will see the picture seems to be much bigger.

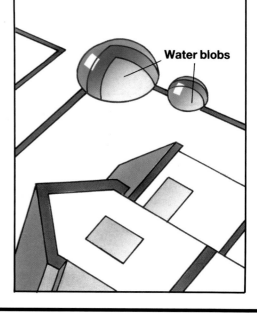

Water blobs

TELESCOPES

People have known for at least 800 years that lenses can magnify objects. The earliest telescopes were made by Flemish lens grinders. These early refracting telescopes (telescopes that use glass lenses to magnify and focus light) did not magnify things very well but an Italian scientist and astronomer named Galileo showed how they could be improved to allow very detailed observations to be made. Reflecting telescopes, invented by the English scientist, Newton, used a concave mirror to collect light from distant objects. As mirrors were easier to make, his reflecting telescopes could be made much larger; which led to further discoveries. Even the radio waves that are produced by distant stars and galaxies can be collected and focused by special dishes that make up radio telescopes. These give valuable information about objects at the very edge of the universe.

Refracting telescopes

A simple refracting telescope can be made with two convex lenses in a tube. More powerful refracting telescopes use many lenses, giving a clearer, sharper image.

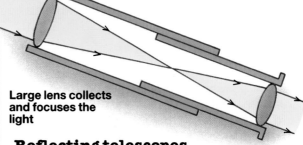

Large lens collects and focuses the light

Reflecting telescopes

Instead of a lens, reflecting telescopes use a concave mirror to collect and focus the light. An eyepiece lens increases the magnification of the image.

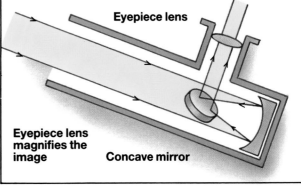

Eyepiece lens

Eyepiece lens magnifies the image

Concave mirror

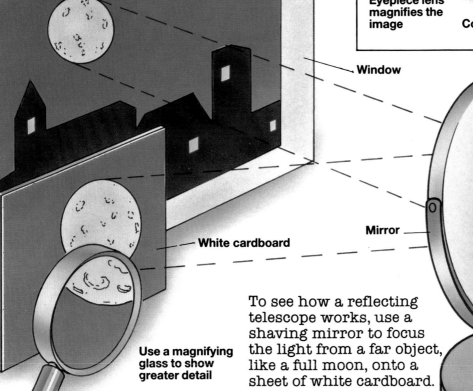

Moon

Window

Mirror

White cardboard

Use a magnifying glass to show greater detail

To see how a reflecting telescope works, use a shaving mirror to focus the light from a far object, like a full moon, onto a sheet of white cardboard.

SPLITTING WHITE LIGHT

White light, the light that comes from the Sun, can be split into seven different colors, called the spectrum of visible light, using a glass prism. A ray of light refracted through the glass is divided up into these seven colors because each color has a different wavelength and so it is bent or refracted through a different angle. The colors are red, orange, yellow, green, blue, indigo and violet.

Sir Isaac Newton

In 1666 Newton discovered that white light is made up of different colors. He used a prism to split a beam of sunlight coming through a slit in a window blind into a darkened room.

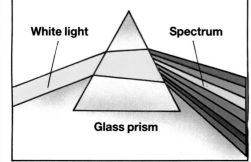

White light   **Spectrum**

Glass prism

Colors of the rainbow

If the Sun is behind you and rain in front of you, the raindrops refract the sunlight, which splits into the spectral colors, forming a rainbow.

Hose pipe rainbow

On a sunny day, adjust the nozzle on a hose pipe to make a fine mist of water. Stand with your back to the Sun and spray the water into the air. You will see a rainbow in the mist of water droplets.

COLOR SPINNERS

Mixing light is very different from mixing paint. The more colors of paints or pigments that are mixed together, the darker the mixture becomes. This is called subtractive color mixing. When light of different colors is mixed, the lighter it becomes. This is called additive color mixing. Light reflected from color spinners mixes in this way. To see this for yourself, use a cup to draw circles on pieces of white stock and carefully cut these out. Push a pencil stub through the center of each spinner disc. (Short pencil stubs are easier to spin.) Color the discs with paints or felt-tipped pens, dividing each disc into seven equal areas and coloring each area with a different color of the spectrum. Then set the discs spinning.

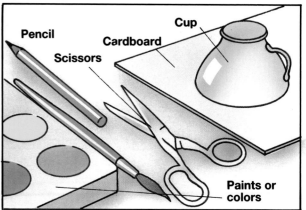

Pencil   **Cup**
Scissors   **Cardboard**
Paints or colors

Use cup to draw circle on cardboard and cut out

MAKE YOUR OWN SPECTRUM

Carefully cut a slit into a sheet of cardboard and tape it to a window. Darken the room by pulling the curtains or blinds, letting through only the ray of light from the slit. Reflect the light ray from a mirror lying in a dish of water onto a piece of white card.

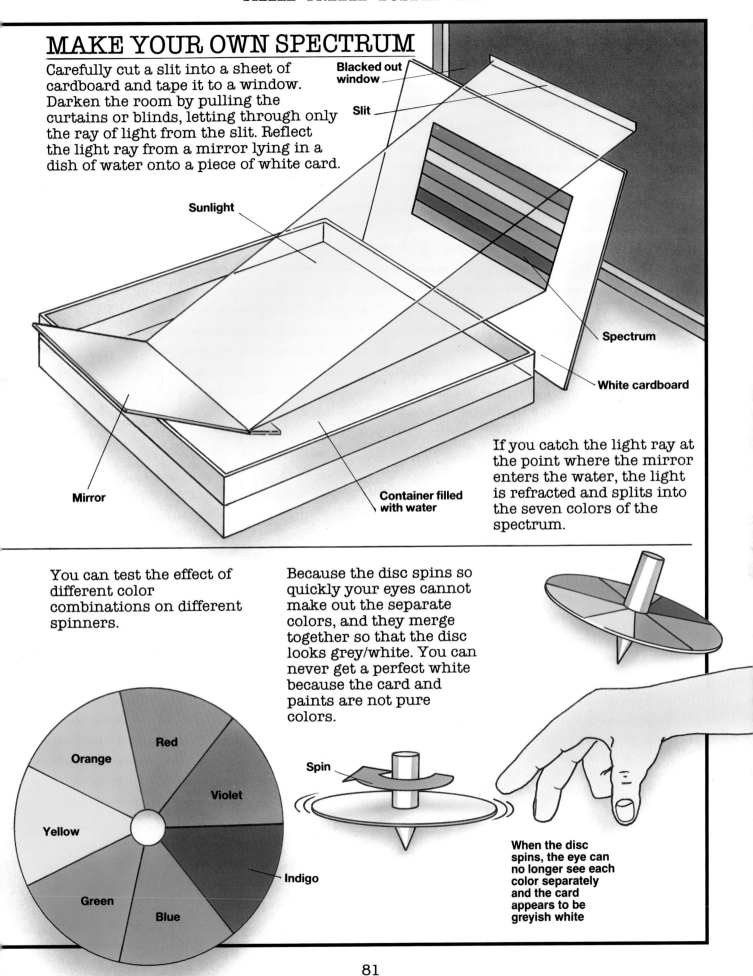

Blacked out window

Slit

Sunlight

Spectrum

White cardboard

Mirror

Container filled with water

If you catch the light ray at the point where the mirror enters the water, the light is refracted and splits into the seven colors of the spectrum.

You can test the effect of different color combinations on different spinners.

Because the disc spins so quickly your eyes cannot make out the separate colors, and they merge together so that the disc looks grey/white. You can never get a perfect white because the card and paints are not pure colors.

Spin

Red

Orange

Violet

Yellow

Indigo

Green

Blue

When the disc spins, the eye can no longer see each color separately and the card appears to be greyish white

Why we see color

We see colors because everything around us absorbs white light and reflects back only some of the spectral colors. White objects reflect almost all the spectrum, but black ones absorb them, reflecting back almost no color at all. Red things absorb spectral colors, reflecting only red light. Blue objects reflect blue light, and so on.

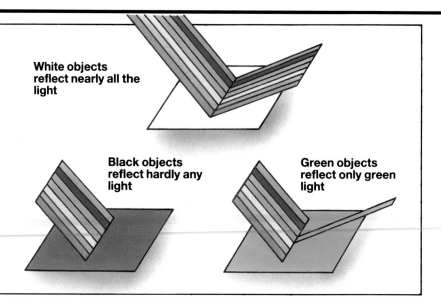

White objects reflect nearly all the light

Black objects reflect hardly any light

Green objects reflect only green light

MIXING WITH LIGHT

You can see what happens when colored light is shone on objects. Ask friends to help you with this experiment. Cover three powerful flashlights with red, blue and green cellophane or color filters. Shine all three flashlights on a sheet of white paper in a very dark room. Look carefully at the way light mixes; it is not like mixing paint. Paints mix to make darker colors, light mixes to make lighter colors. Notice how the color of objects like your toys changes when you shine colored light on them.

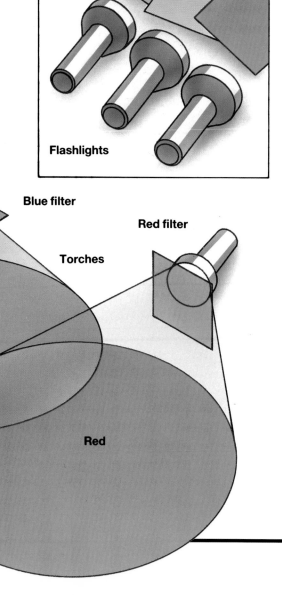

Colored cellophane

Flashlights

Green filter

Where all the lights meet, white light is produced

Blue filter

Red filter

Torches

Blue

White

Green

Red

SEPARATING COLORS

Many colors, like food dyes and the ink in felt-tipped pens, are mixtures of colors. It is possible to split some colors by a process called chromatography. Put a large spot of one color at the bottom of a strip of blotting paper. Tape the strip to a pencil and rest it on a jar with some water in it. The water soaks into the blotting paper, carrying the ink with it. The different colors in the ink move at different speeds, so they spread out along the paper.

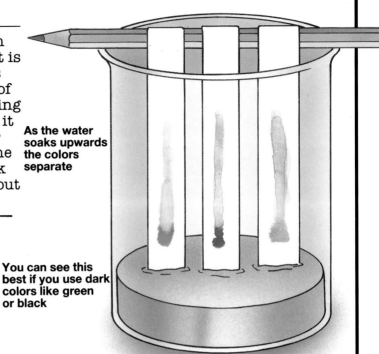

As the water soaks upwards the colors separate

You can see this best if you use dark colors like green or black

Strips of blotting paper

Glass

Water

Pencil

Felt-tipped pens

SEEING RED

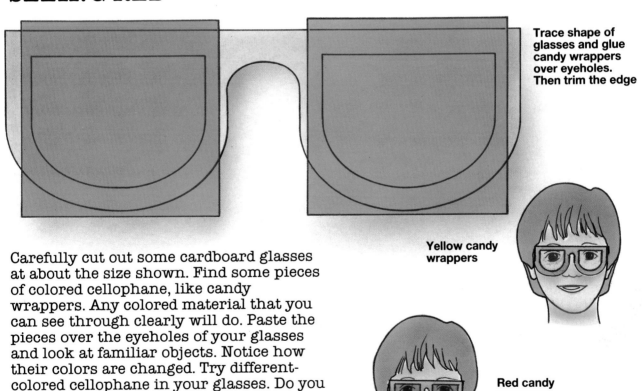

Trace shape of glasses and glue candy wrappers over eyeholes. Then trim the edge

Yellow candy wrappers

Red candy wrappers

Carefully cut out some cardboard glasses at about the size shown. Find some pieces of colored cellophane, like candy wrappers. Any colored material that you can see through clearly will do. Paste the pieces over the eyeholes of your glasses and look at familiar objects. Notice how their colors are changed. Try different-colored cellophane in your glasses. Do you like a yellow world or would you prefer a cool blue one? Do some colors make you feel warm, do others make you happy? Which is your favorite color?

CHEMICAL ENERGY

Chemical energy is a form of stored energy or fuel. Most chemical energy is released by burning, giving heat and light. Fuel like oil, gas and coal was formed over millions of years from dead plants and animals. Food is also stored energy. Plants make their own food from sunlight by a process called photosynthesis. Some animals (herbivores) eat plants to get the energy they need to grow. Other animals (carnivores) get their food energy by eating other animals.

Why do we eat?
We eat to get energy. Food is broken down inside our bodies into proteins, vitamins, minerals and so on. This is digestion. It provides all the things we need to keep warm, to grow and to stay healthy.

Solar energy
The Sun gives us all the energy we need, but we cannot live off warmth and sunlight alone. We need food. The Sun makes our food, which gives us our energy.

PRINCIPLES OF CHEMICAL ENERGY

Food chains

Herbivores are animals that eat plants to get the energy they need to live. Other animals, the carnivores, eat the herbivores. When animals die, scavengers such as insects eat the dead tissue. Dead plant material, such as leaf mould and tree bark, forms the food of many fungi. Decayed plant and animal tissue produces minerals to fertilize the land for more plants to grow. Everywhere there are complex food chains of plants and animals living off each other. Each link in the chain gets the energy it needs for life from other plants or animals. Every food chain begins with green plants and the Sun.

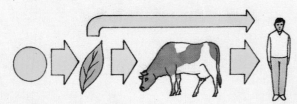

The food chain

Oxygen and carbon dioxide

The chemical energy in food that is essential for life can only be released in the body's cells if oxygen is present there as well. All animals need a constant supply of oxygen, breathed in from the air, so that a complex series of chemical reactions can take place to release the energy. Animals breathe out carbon dioxide as a waste product (something they do not need) of these reactions. Plants take in carbon dioxide, from the air or from water, to create their own energy. They give out oxygen as a product of photosynthesis, putting the oxygen animals need back into the atmosphere.

Plants take in carbon dioxide and give out oxygen

People and animals breathe in oxygen and breathe out carbon dioxide

Animals and plants

Animals depend upon plants for their food energy. Animals are energy-consumers; plants are energy-producers, but where do plants get their energy supply?

Photosynthesis

Green plants make energy through a process called photosynthesis. Complex chemical reactions turn carbon dioxide, water and sunlight into energy-rich sugars and oxygen, using a chemical, chlorophyll, that gives leaves their green color.

CANDLE SNUFFER

A candle burns until its wax has gone, unless it is blown out. It gets its chemical energy from the wax. But it also needs air to make it burn. Put a candle in a tray of sand or soil and ask an adult to light it for you. Put a large glass (not plastic) jar over the candle. When the candle has used up the air in the jar it goes out.

Place the jar over the candle and the flame soon goes out

Jar

Sand

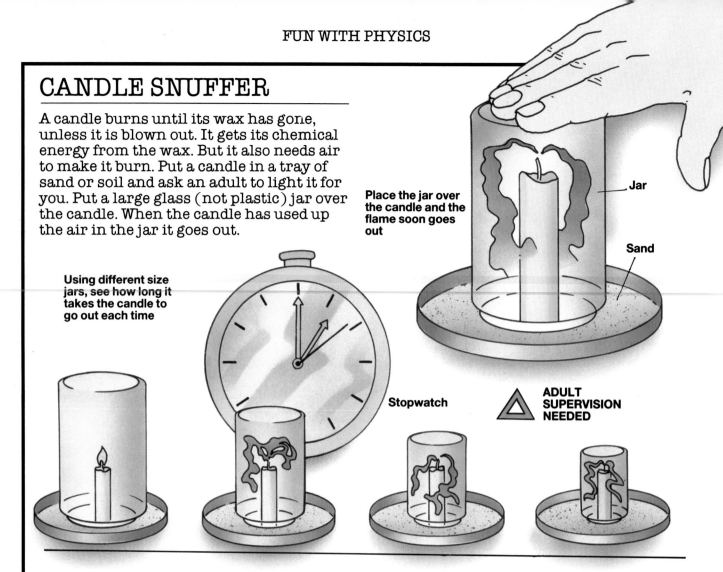

Using different size jars, see how long it takes the candle to go out each time

Stopwatch

ADULT SUPERVISION NEEDED

DOES ALL AIR BURN?

Put a candle in a dish of water. Ask an adult to light it. Put a glass jar over it, resting on some coins to keep a gap between the jar and dish. As the candle burns, it uses up part of the air in the jar. The space left by the used up air is filled by water sucked in from the dish, but it only fills about a fifth of the space in the jar.

Jar

Candle

Coins

Dish with water

ADULT SUPERVISION NEEDED

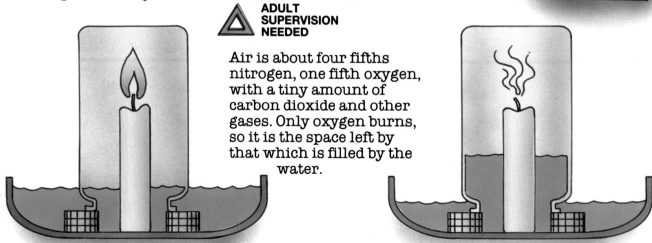

Air is about four fifths nitrogen, one fifth oxygen, with a tiny amount of carbon dioxide and other gases. Only oxygen burns, so it is the space left by that which is filled by the water.

FIRE-FIGHTING

To put out a fire you must stop oxygen from reaching it. Water will drench the fire, so air can't reach it, fire-proof blankets can be thrown over small fires, and foam or carbon dioxide extinguishers will smother a fire in the same way. You can make your own carbon dioxide extinguisher with bicarbonate of soda (baking powder) and vinegar. The baking soda reacts with the acid in vinegar to make carbon dioxide gas. Ask an adult to help you set up a small candle in a sand tray and mix the ingredients in a large beaker. As the mixture froths, hold the beaker above the flame. **DON'T** pour the liquid on the flame, just let the gas flow over the flame.

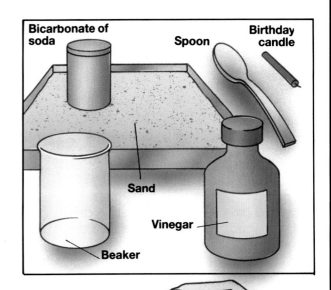

Bicarbonate of soda

Spoon

Birthday candle

Sand

Vinegar

Beaker

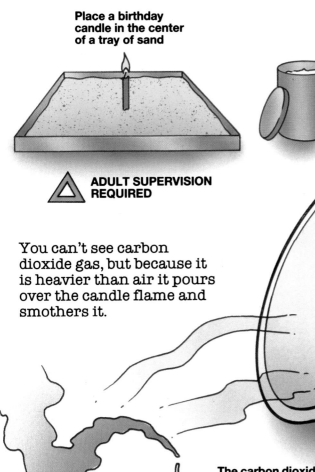

Place a birthday candle in the center of a tray of sand

Add a large spoonful of bicarbonate to the beaker of vinegar

ADULT SUPERVISION REQUIRED

You can't see carbon dioxide gas, but because it is heavier than air it pours over the candle flame and smothers it.

The carbon dioxide produced by the reaction drifts downwards and puts out the flame

87

GROWING FOOD

Compare the effects of growing seeds under different conditions. Use fast-growing seeds, such as cress, and prepare your samples in exactly the same way. Line the trays with a pad of cotton and soak each with the same amount of water, so that it is very damp but not running with excess water. Sprinkle the same quantity of seeds evenly over the cotton. Keep one in a warm, sunny position – this is called the control sample. Place a second sample in the same place, but cover the tray with a lid that does not let in light. When the cress in the first tray is growing well, compare the growth with the second sample. What differences do you notice?

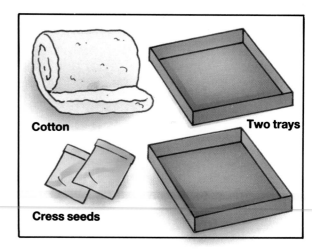

Cotton　　**Two trays**

Cress seeds

Fill both trays with cotton and soak with water

Spread the seed out evenly

With other samples you can test the effect of pollution on plant growth. Grow one sample using pure water and place it in a warm, sunny position as before. Make up a second tray, but ask an adult to soak the cotton in water containing two tablespoons full of bleach. Add a similar amount of detergent to a third sample. Compare the growth and appearance of the cress plants. What differences do you notice?

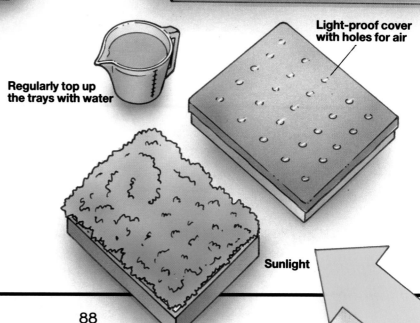

Regularly top up the trays with water

Light-proof cover with holes for air

Sunlight

ROTTING

Find several glass, screw-topped jars and place a different sample of food or even some milk into each one. Screw on the lids and seal with sticky tape. Watch the changes that take place over a few weeks. Bacteria in the food causes decay. Because the substances that form may be dangerous, **DO NOT OPEN** the jars. Dispose of them carefully.

Food

Four sealed jars

Apple

Bread

Banana

Milk

RUSTING

Some metals that are left unprotected may begin to rust. Use samples of steel wool or clean iron nails or even straight pins to see why this happens. Leave one piece of steel wool or some nails or pins in a plastic dish with a little water. Keep the sample damp. In another dish completely cover the wool or nails in water from a kettle that has boiled and then gone cold. Leave a third sample in a dry dish in a warm, dry spot. What do you predict will happen to your metal samples?

Three bowls

Water

Steel wool

To rust, metals need moisture and oxygen. Boiled water has no air, so the sample may not rust. Samples kept damp and open to the air will rust quickly.

Small amount of water to keep the steel wool damp

Keep this bowl dry

Fill the bowl with water from the kettle so that the steel wool is completely covered

PHYSICS OF HEAT

Heat is a form of energy. Atoms and molecules, the tiny particles that make up everything that exists, are constantly moving and the amount of heat within any substance depends on how much and how fast its atoms move. When more heat energy is given to something its temperature rises, it expands (becomes larger) and it may change from a solid to a liquid or from a liquid to a gas. Heat can be transferred from one place to another in three ways, conduction, convection or radiation, but it can only move from a hotter area to one that has a lower temperature.

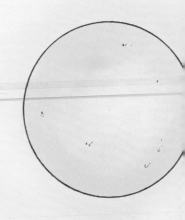

Convection
Hot air rises. The air over a fire, or over an area heated by the Sun, gets hotter and rises, letting colder air flow in below it. Heated water also rises and cooler water flows in to form swirling currents.

Radiation
Nearly all our heat and light come from the Sun. The Sun is 93 million miles away, but its heat and light radiation pass through space and reach us at the speed of light eight minutes later.

Conduction
Heat moves through solid objects by conduction. Put a metal spoon in a hot drink and the handle gets warm. The heat around the spoon warms the metal and passes through it along the handle. Some things, like metals, are good heat conductors. Others, called insulators, are not.

PRINCIPLES OF TEMPERATURE

How hot is hot?

Get three bowls with water at different temperatures. Put one hand into a bowl of hot water and the other into cold water. After half a minute put both hands into a bowl of warm water. This water should feel the same to both hands but it doesn't. The cold hand thinks it is in hot water, the hot hand thinks it is in cold water

Hot   Warm   Cold

Calibration

Measuring temperatures by touch would mean many mistakes, so we need a standard measure. Liquids, like alcohol and mercury, expand when heated. A thin tube filled with one of these liquids reacts to temperature changes. The tube can be marked at points showing where the liquid expands to when it is heated, and where it contracts to when cooled. The standard temperature points are fixed at the freezing point and boiling point of pure water. The thermometer tube can then be marked at intervals between these fixed points so that it can be used to measure temperatures accurately. Marking a thermometer in this way is called calibration.

Temperature scales

There are three temperature scales commonly used today. Gabriel Fahrenheit of Germany fixed the freezing point at 32°F and boiling point at 212°F. Anders Celsius of Sweden devised the Celsius scale with 0°C as the freezing point and 100°C as the boiling point. Later a British scientist, William Kelvin, gave us the idea of an absolute zero, the lowest possible temperature, which sets the freezing point of water at 273°K and the boiling point at 373°K.

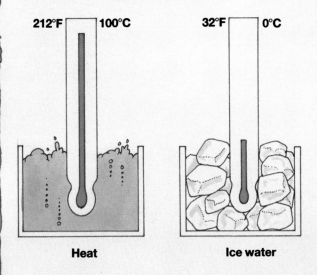

Heat   Ice water

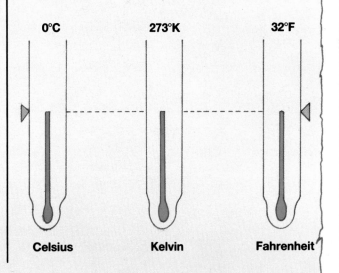

Celsius   Kelvin   Fahrenheit

KEEPING WARM

Find four jars with lids and ask an adult to pour hot water of the same temperature into each one. Replace the lids, wrap one jar in a handkerchief, another in a thick wool scarf and put a third in a box of polystyrene chips or sawdust. Don't wrap the fourth jar. Record the temperature in each jar with a thermometer every half hour. Which one stays warmest and which one cools fastest?

Thermos flask

A thermos flask has a sealed double-walled glass container with a vacuum inside. Heat cannot be conducted or convected through a vacuum. Radiated heat is reflected back in by the silvered glass that acts like a mirror. The only heat loss is through the cap, which is well insulated, so hot liquids stay hot for a long time.

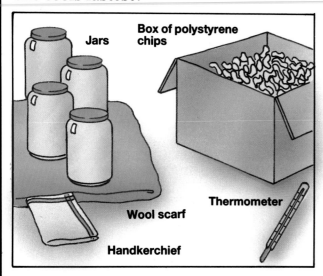

Jars

Box of polystyrene chips

Wool scarf

Thermometer

Handkerchief

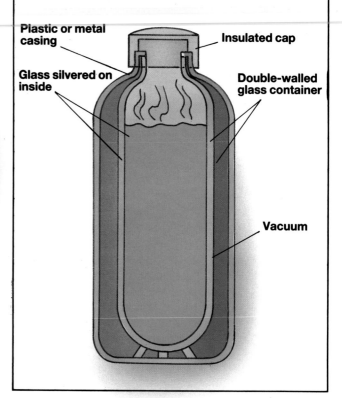

Plastic or metal casing

Insulated cap

Glass silvered on inside

Double-walled glass container

Vacuum

A well wrapped (insulated) jar keeps its heat best. Use this test to compare all kinds of materials to find the best insulators.

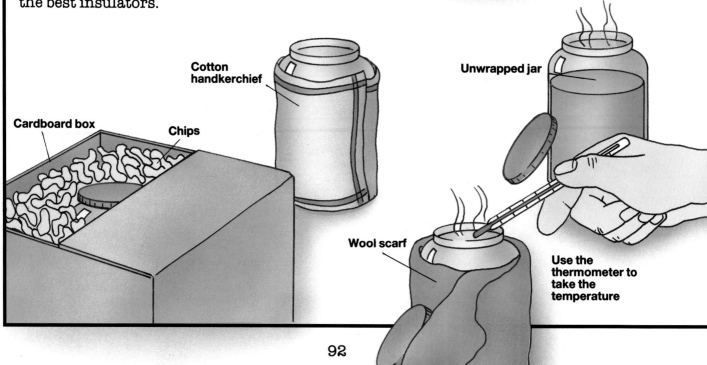

Cotton handkerchief

Unwrapped jar

Cardboard box

Chips

Wool scarf

Use the thermometer to take the temperature

RAISING YOUR CAP

Water is at its most dense, the point where it takes up the smallest amount of space, at a temperature of 39°F. As water freezes into ice it actually expands. This is why water pipes can burst in the winter. Fill a plastic bottle to the very top with cold water. Cover the top with a foil cap and place the bottle upright in a freezer. Leave overnight, then see what has happened. As the water freezes it can only go up.

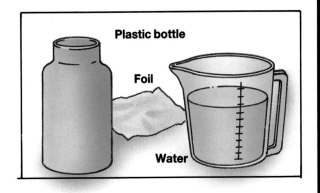

Plastic bottle

Foil

Water

When the bottle is full, put a foil cap on and place it in the freezer

As it freezes the water will expand

Ice takes up more space than water so it eventually pushes the cap off

Foil cap

PUDDLE WATCHING

After a period of heavy rain, find a large puddle and scratch around its edge with a sharp stone, or mark it with chalk. Leave the puddle for about 15 minutes and do the same again. Repeat this four or five times. What do you notice about the puddle? The water seems to be disappearing. But where is it going? Can you estimate how much water disappears during each 15-minute period?

Sun

What is happening is that heat from the Sun, helped by wind, makes the water evaporate. You cannot see water vapor in the air, but it is carried into the atmosphere, forming clouds.

Evaporation

Draw a new line every 15 minutes

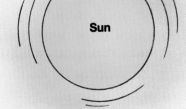

Chalk lines

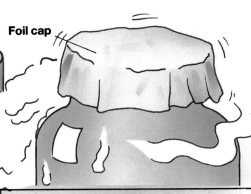

CONVECTION CURRENTS

Put a few drops of ink or food dye in a bottle and fill it with hot water. Cover the top with a foil cap with some thread attached. Lower the bottle into a tank full of clear, cold water. Let the water settle, then pull the foil away, using the thread. Watch as the hot colored water rises in the tank and swirls around, following the convection currents of cold water flowing in towards the bottle.

Seal with silver foil attached to thread

Hot water plus colored ink

The hot colored water circulates, following the convection currents

Cold water

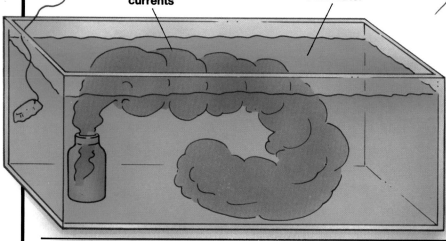

Hot water

Plastic bottle

Dye

Thread

Silver paper

HEAT CONDUCTORS

You can watch the speed with which heat is conducted along a solid object by using candle wax. Ask an adult to help you drip candle wax onto a long nail. Hold the nail with a clothespin or pair of pliers. Put the end of the nail in a candle flame. As the heat is conducted along the nail the wax begins to melt and drip off.

Using oven gloves, drip candle wax on to a nail and let it cool

⚠ **ADULT SUPERVISION NECESSARY**

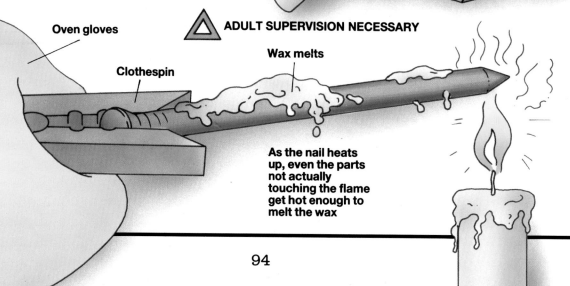

Oven gloves

Clothespin

Wax melts

As the nail heats up, even the parts not actually touching the flame get hot enough to melt the wax

SPIRAL SNAKE

Carefully cut a circle from a sheet of thin paper and cut it into a spiral shape. Hang the spiral from a length of thread over a hot radiator or storage heater. **DO NOT** hang it over a fire or flame. The rising hot air currents catch the spiral, making it twist and turn. In a room, the convection currents that form around a radiator or fire carry the heat all around the room, so that eventually every corner of the room is warmed.

Hang the snake from a string

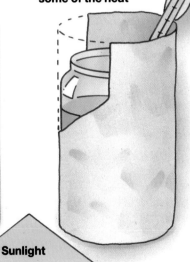

Draw and decorate the spiral snake on card and then cut it out

Hot air rises and the snake twists and turns

RADIATED HEAT

Radiated heat can be reflected away. Take two identical glass jars, filled with the same amount of cold water. Cover one in black paper and the other in silver foil, then put them in a sunny place for about an hour. Check the temperature. What do you find?

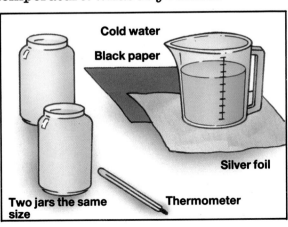

Cold water

Black paper

Silver foil

Two jars the same size

Thermometer

Black paper absorbs heat and the temperature rises

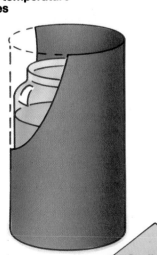

Take the temperature in each jar after one hour

Silver foil reflects some of the heat

Sunlight

95

SOURCES OF ENERGY

All of our energy comes from the Sun, whether it is direct heat and light from the Sun's radiated energy, or stored chemical energy in fuels and food. Our problem is that our demand for energy increases every day. Everything in our lives depends on the constant supply of power, particularly electrical power. Think of all the things you use that need electricity to work. Almost every industry needs electrical energy. We use vast quantities of energy. Without it, life as we know it could not continue. So where can all this energy come from?

Gas, coal and oil
We think of fuels as gas, coal and oil. Coal has been important for hundreds of years and oil provides us with petroleum and plastics that are in increasing demand. What can we do when these fuels run out?

Wind and water
Waves build up energy as they move across oceans and wind can be powerful enough to uproot trees. Engineers are investigating many interesting ways to harness this natural power, including wave-driven turbines and wind generators.

Nuclear power

Scientists discovered that elements such as uranium and plutonium can be used as a source of nuclear energy, giving out vast quantities of heat that could be used destructively in bombs or as a controlled source of energy in nuclear power stations.

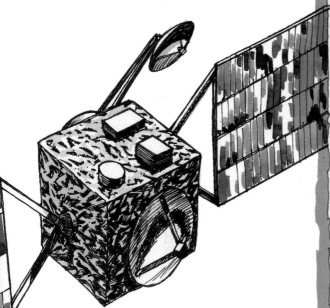

Solar power

In many hot countries, solar panels are used to capture the Sun's heat energy. Water is pumped through a series of pipes laid under a glass panel which is heated by the Sun, giving cheap hot water.

PRINCIPLES OF SUPPLY

Fossil fuels

Fossil fuels were made over millions of years by the action of pressure and heat upon layers of dead plant and animal tissue under the ground and the sea. Because this process takes so much time, and because we use so much of these fuels to make electricity, Earth's reserves of fossil fuels have almost gone, so we need to develop other sources of power.

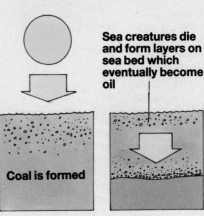
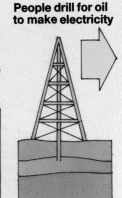

People drill for oil to make electricity

Sea creatures die and form layers on sea bed which eventually become oil

Coal is formed

Renewable power

Wind always blows and water constantly flows, providing a free, continuous source of energy. We need to find ways to use this. One that has been used for many years is hydro-electricity. Water trapped behind a dam is released and rushes over turbines, turning them very fast. The turbines are connected to generators that produce electricity.

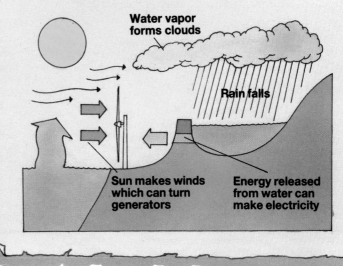

Water vapor forms clouds

Rain falls

Sun makes winds which can turn generators

Energy released from water can make electricity

MAKE YOUR OWN WINDMILL

Modern windmills have two or three blades that are specially shaped like an aircraft's wing (an airfoil shape) to catch the wind and turn at the highest possible speed. The energy that is produced when the blades turn is used to turn a generator that makes electricity. A large wind generator can produce a great deal of electrical power. The windmill shown here turns a little acrobat figure attached to a crank on the windmill blades. Follow the step-by-step instructions to set your acrobat tumbling.

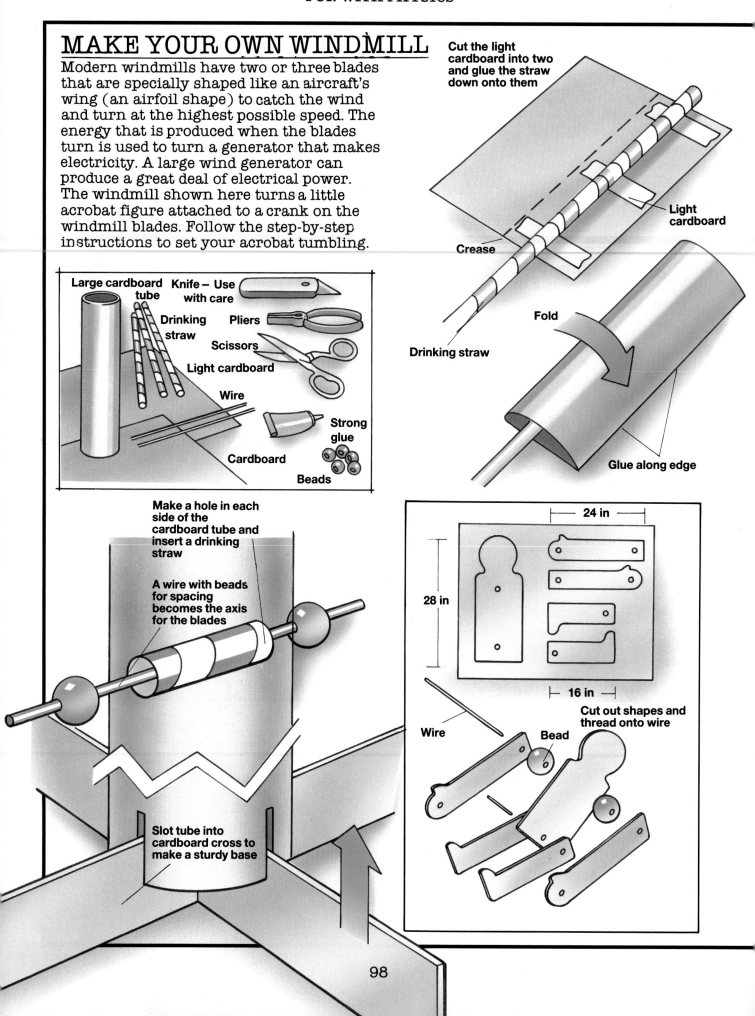

Cut the light cardboard into two and glue the straw down onto them

Crease

Light cardboard

Drinking straw

Fold

Glue along edge

Large cardboard tube

Knife – Use with care

Drinking straw

Pliers

Scissors

Light cardboard

Wire

Strong glue

Cardboard

Beads

Make a hole in each side of the cardboard tube and insert a drinking straw

A wire with beads for spacing becomes the axis for the blades

Slot tube into cardboard cross to make a sturdy base

24 in

28 in

16 in

Wire

Bead

Cut out shapes and thread onto wire

FUTURE ENERGY

Burning fossil fuels can damage our environment by releasing harmful gases. Carbon dioxide released from burning fuels forms a layer in the atmosphere that causes global warming. We need to get more energy from the Sun, wind and waves. We also need to save energy by recycling household waste like glass, metals and paper.

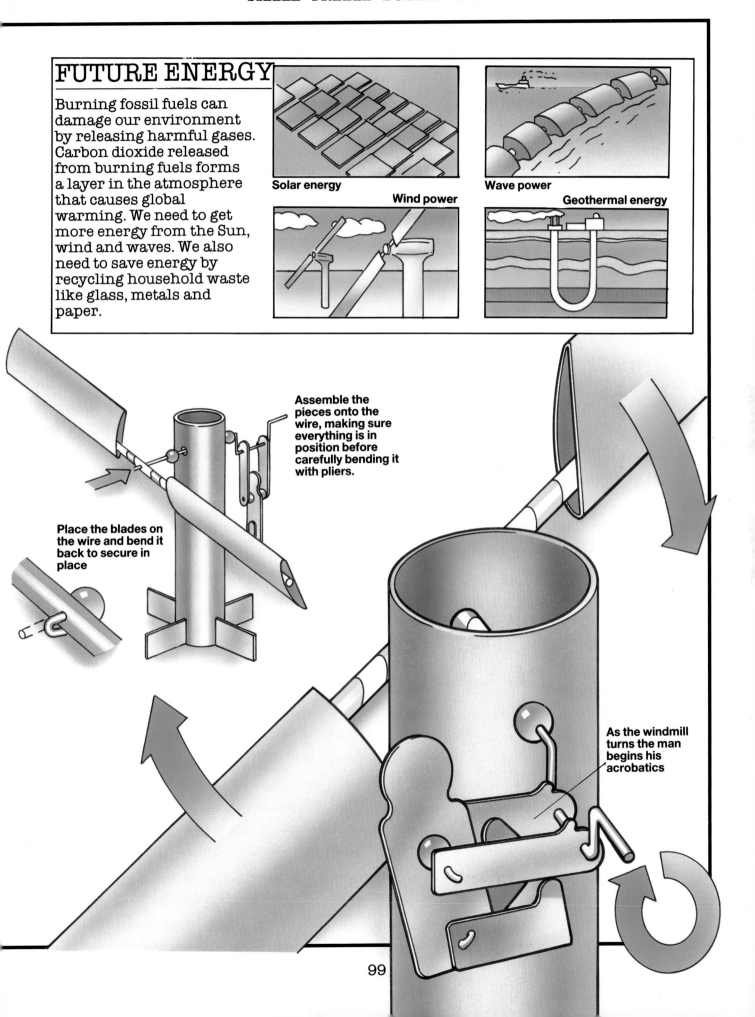

Solar energy

Wave power

Wind power

Geothermal energy

Assemble the pieces onto the wire, making sure everything is in position before carefully bending it with pliers.

Place the blades on the wire and bend it back to secure in place

As the windmill turns the man begins his acrobatics

NUCLEAR FISSION

Everything is made from tiny particles called atoms. Some atoms are so large they are unstable and cannot hold together. They break up, giving out a burst of energy and releasing nuclear particles called neutrons. In a large mass of unstable atoms these "free" neutrons can make more atoms break up, creating a continuous reaction called nuclear fission. Early this century it was discovered that some substances like uranium could be made to break up in this way, giving out heat energy that can be used to make electricity. The problem is that this reaction is very dangerous and has to be very carefully controlled. It also leaves large amounts of deadly radioactive waste to be disposed of.

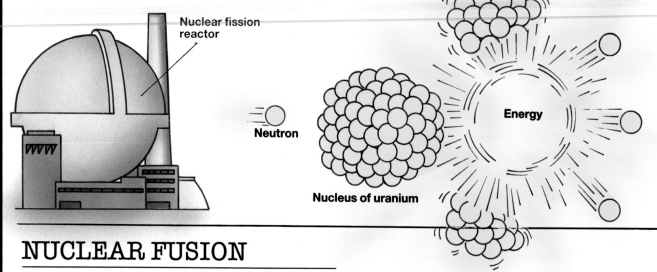

Nuclear fission reactor

Neutron

Nucleus of uranium

Energy

NUCLEAR FUSION

The Sun's energy comes from a process called nuclear fusion. Under very strong heat and pressure, special hydrogen atoms, called deuterium and tritium, combine together to form helium. As this happens, tremendous amounts of heat and light energy are produced. This principle has been used to make nuclear fusion weapons (often called hydrogen bombs) that have terrifying destructive power. Scientists all over the world are trying to find a way to control fusion reactions here on Earth. If they succeed, the energy produced will last us almost for ever. Also, it will not bring any of the current problems that we have with waste polluting the environment, just cheap energy for everyone.

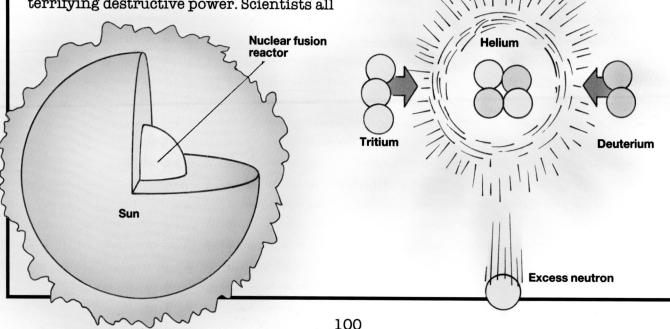

Nuclear fusion reactor

Sun

Helium

Tritium

Deuterium

Excess neutron

NOTES FOR PARENTS AND TEACHERS

Science education should enable children to learn about themselves and the world around them through firsthand experience and practical investigation. Science can be thought of as a body of knowledge or as a method of enquiry, but it is only by combining the two that a true understanding can develop.

This science book introduces the essential background knowledge component, but also offers many opportunities to apply this knowledge through exciting investigations and experiments.

101 Physics Tricks is concerned principally with the science of energy and with forces that also form the foundation of many technological developments.

Section 1, "Puff, Squeeze, Bang, Blow", addresses the physics of natural forces. Children can discover:
- that weight is a force, measured in Newtons
- that a parachute can help to overcome the effect of gravity
- the effects of friction that make movement possible, but also waste energy and wear out machine parts
- that energy can be stored in stretched and twisted elastic materials and be used to drive simple machines
- the secrets of flight, including the airfoil principle and controlling movement through the air
- the uses of natural energy such as wind and water power
- how we hear sounds and the effects of sound energy

Section 2, "Click, Flash, Buzz, Whirr", is concerned with magnetism and electricity. In this section children discover:
- how to make their own compass
- the attractive and repellent properties of magnets
- the power of magnetism to pass through solid objects
- the principles of static electricity and how static can stick balloons to the ceiling
- how electrical power is generated using Faraday's discoveries of electron flow in coils
- how to construct a simple circuit, lighting two or more bulbs from one power source using parallel and series circuits
- that conductors and insulators can be identified with a simple test
- the effects of electromagnetism and how to wind your own electromagnetic coil

Section 3, "Sizzle, Freeze, Bubble, Glow", covers important aspects of light energy, chemical energy and heat.
Children discover:
- the importance of the Sun's energy to life on Earth
- the way in which light can be reflected and refracted
- the qualities of materials showing transparency and opacity, and the way in which shadows are cast
- how eyes react to light and color and the effect of color mixing and light filters
- that white light can be split into the seven spectral colors which are a narrow band of electromagnetic radiation
- the link between sundials and shadow puppets
- how lenses and mirrors can improve our vision and create fascinating effects
- the importance of stored energy in food and the principle of food chains, beginning with sunlight and green plants
- the relationship between heat and temperature and the ways in which heat can be transferred through conduction, convection and radiation
- the facts about alternative energy and the pursuit of infinite energy sources through nuclear fusion.

INDEX